How Big and Still Beautiful? Macro-Engineering Revisited

AAAS Selected Symposia Series

Published by Westview Press
5500 Central Avenue, Boulder, Colorado

for the

American Association for the Advancement of Science
1776 Massachusetts Avenue, N.W., Washington, D.C.

How Big and Still Beautiful? Macro-Engineering Revisited

Edited by Frank P. Davidson,
C. Lawrence Meador and Robert Salkeld

Routledge
Taylor & Francis Group
LONDON AND NEW YORK

First published 1980 by Westview Press

Published 2018 by Routledge
52 Vanderbilt Avenue, New York, NY 10017
2 Park Square, Milton Park, Abingdon, Oxon OX14 4RN

Routledge is an imprint of the Taylor & Francis Group, an informa business

Library of Congress Cataloging in Publication Data
Main entry under title:
How big and still beautiful? macro-engineering revisited
(AAAS selected symposium ; 40)
Papers presented at a symposium held at the 1979 annual meeting of the American Association for the Advancement of Science and sponsored by the Section on Engineering and the Section on Industrial Science of the AAAS, and the American Institute of Aeronautics and Astronautics.
Bibliography: p.
Includes index.
1. Engineering--Congresses. I. Davidson, Frank Paul, 1918– II. Meador, C. Lawrence III. Salkeld, Robert, 1932– IV. American Association for the Advancement of Science. Section on Engineering. V. American Association for the Advancement of Science. Section on Industrial Science. VI. American Institute of Aeronautics and Astronautics. VII. Series: American Association for the Advancement of Science. AAAS selected symposium ; 40.
TA5.H67 620'.004'2 79-23498
ISBN 0-89158-792-6

ISBN 13: 978-0-367-02209-9 (hbk)

About the Book

This volume examines fundamental issues of macro-engineering--now a planetary norm--from the viewpoints of psychiatry, social science, management, and law. The contributors suggest a general theory to guide future decisions on large-scale projects and programs and analyze specific cases in the context of a set of public-interest guidelines.

About the Series

The *AAAS Selected Symposia Series* was begun in 1977 to provide a means for more permanently recording and more widely disseminating some of the valuable material which is discussed at the AAAS Annual National Meetings. The volumes in this *Series* are based on symposia held at the Meetings which address topics of current and continuing significance, both within and among the sciences, and in the areas in which science and technology impact on public policy. The *Series* format is designed to provide for rapid dissemination of information, so the papers are not typeset but are reproduced directly from the camera-copy submitted by the authors, without copy editing. The papers are organized and edited by the symposium arrangers who then become the editors of the various volumes. Most papers published in this *Series* are original contributions which have not been previously published, although in some cases additional papers from other sources have been added by an editor to provide a more comprehensive view of a particular topic. Symposia may be reports of new research or reviews of established work, particularly work of an interdisciplinary nature, since the AAAS Annual Meetings typically embrace the full range of the sciences and their societal implications.

WILLIAM D. CAREY
Executive Officer
American Association for
the Advancement of Science

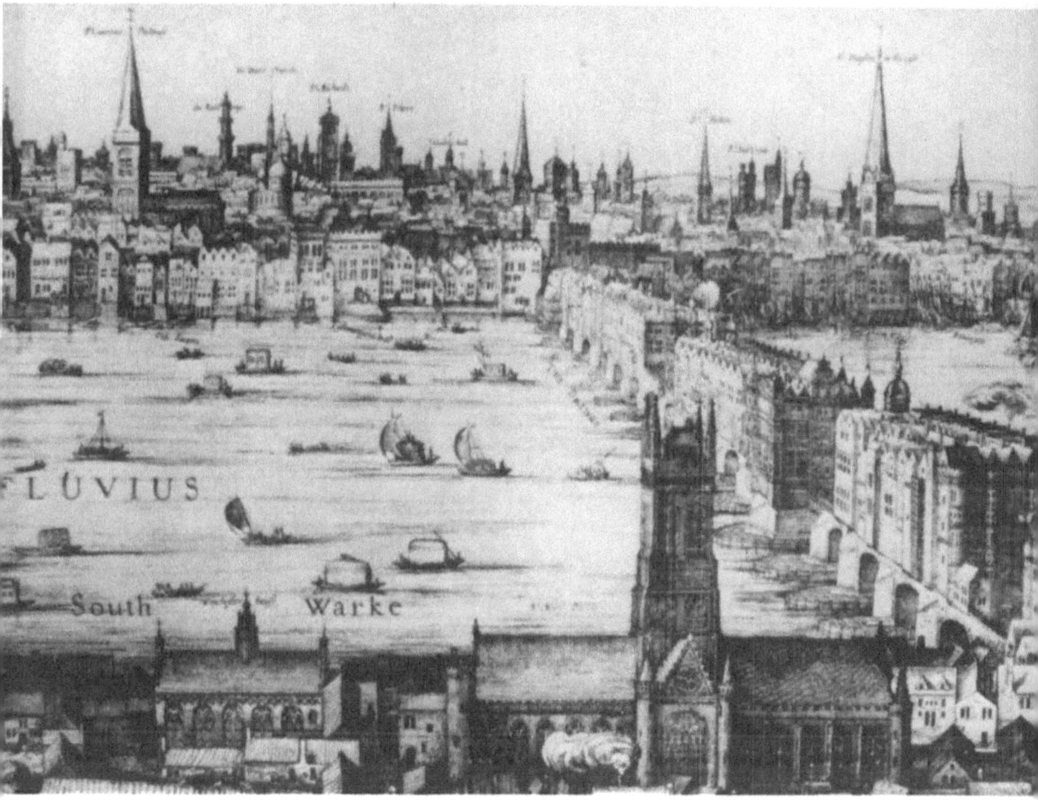

"London Bridge and the City Skyline Beyond." From a lithograph by Wenceslaus Hollar, c. 1620. Courtesy of the Trustees of the British Museum.

Contents

Figures, Tables and Illustrations

About the Editors and Authors

Frank P. Davidson, *chairman of the System Dynamics Steering Committee, Alfred P. Sloan School of Management, Massachusetts Institute of Technology, holds a J.D. from Harvard Law School, but has been involved in macro-engineering for the past 20 years. He has been advisor on projects of unusual magnitude and complexity to the U.S., Canadian and British governments and to corporations and universities. He was founding president of the Institute for the Future in California and founder of the Channel Tunnel Study Group (planning group for the English Channel tunnel) and is a member of its governing board. He is also president of Technical Studies, Inc., the U.S. participant in the Channel Tunnel Study Group, and is vice-chairman of the Institute for Educational Services in Massachusetts. Among his many publications is* Macro-Engineering and the Infrastructure of Tomorrow *(AAAS Selected Symposium No. 23, Westview, 1978), of which he was senior editor.*

C. Lawrence Meador, *lecturer in the School of Engineering at Massachusetts Institute of Technology, is a specialist in computer-based information and decision support systems for engineering design, policy analysis, planning, and control. He is president of Decision Support Technology, Inc., and has served as consultant to the U.S. Departments of Energy and Defense and several major corporations. He has been assistant director, research associate and instructor for the Center for Information Systems Research, Alfred P. Sloan School of Management, and was a founding member of the Clinical Decision Making Group at MIT's Project Mac. He has published widely, especially on decision support systems, and is vice-chairman of the Eastern Hemisphere and Latin America Area Committee of the Institute for Electrical and Electronic Engineers' Computer Society. He also serves as an editor of Computer Communications.*

Robert Salkeld *is a consultant on aerospace and military planning to industrial and government organizations and chairman of the Technical Committee on Space Systems for the American Institute of Aeronautics and Astronautics. He participated in early studies of reusable launch vehicles and spacecraft and of space stations, and worked on Pioneer I, the first deep space probe. In addition to holding numerous patents on space systems devices, he is the author of more than 30 publications, including* **War and Space** *(Prentice-Hall, 1970). He is a fellow of several professional societies.*

Ramón Barquín, *manager for External Programs for IBM Americas/Far East Corporation and research fellow at MIT, is a specialist in technology transfer. He is the author of publications on transfer of computer technology, creation of technology in Latin America, and cultural differences and information systems utilization, and is chairman of the International Chapters Area, IEEE Computer Society.*

Kenneth W. Billman *is group leader of the Advanced Projects Group in the Materials and Physical Sciences Branch at NASA's Ames Research Center, specializing in space power concepts, advanced energy systems and laser energy conversion. He is the author of numerous publications on these topics, and is the editor of* **Radiation Conversion in Space,** *Volume 61 of the AIAA Progress in Astronautics and Aeronautics Series. He has been chairman of three national conferences on radiation energy conversion sponsored by NASA, and he holds three patents on his work in these fields.*

Stuart W. Bowen *is an aerospace engineer with Beam Engineering in California. His past areas of research have included design development and operation of DC arc plasma jets, quantitative plasma diagnostics, kinetics of gas and plasma reactions, and modeling and calculating recombination and excited state distributions in plasma flows and plasmadynamic lasers. His current research is concerned with calculation of characteristics of insolation on ground sites from a system of orbiting mirrors (SOLARES) and the design of the reflecting mirrors.*

John P. Craven *is currently Hawaii State Marine Affairs Coordinator, dean of Marine Programs and professor of Ocean Engineering at the University of Hawaii, and director of the Law of the Sea Institute. He has published widely in the fields of marine hydrodynamics and technology, as well as on arms control and the law of the sea. He is a member of the advisory committee to the U.S. delegation to the Third United Nations Law of the Sea Conference, and is currently a member* *of the National Academy of Engineering, serving on its Marine Board and Ocean Policy Committee. He has received numerous awards and honors for his work, including Distinguished Civilian Service Awards from the Department of Defense and the Navy.*

William P. Gilbreath, *a research scientist at Ames Research Center, is currently working on concept development and feasibility of large projects such as the generation, storage and transmission of large amounts of energy in space. He has carried out studies of the interaction of aerospace materials with various environments, including sea water, rocket propellants, high vacuum and various gaseous species. He holds memberships in various professional and honorary societies and has published a number of papers in his fields of interest.*

Cynthia Oudejans Harris *is a psychiatrist for the Gestalt Institute of Cleveland; a faculty member at Case Western Reserve University, Department of Community Health; and a consultant psychiatrist to the Ohio state hospital system. Among her many interests is a concern for the impact of the modern architectural environment on the people who must live in it.*

Ewald Heer, *director of Autonomous Systems Technology at the Jet Propulsion Laboratory, Pasadena, is also an adjunct professor of industrial and systems engineering and director of the Institute for Technoeconomic Systems at the University of Southern California. A specialist in development of automated and space systems, he played a leading role in the planning of the NASA robotics program. He has organized and chaired two international conferences on remotely controlled and automated systems; has written many papers and has edited two books in these fields.*

Hans R. Huessy, *professor of psychiatry at the University of Vermont College of Medicine, is a specialist in child and community psychiatry and is concerned with the social impact of macro-engineering projects. He has published articles on social organization and mental health and is the author of* **Mental Health with Limited Resources** *(Grune and Stratton, 1966).*

Cordell W. Hull, *vice-president and treasurer for the Bechtel Corporation, has degrees in both engineering and law, and has extensive experience in international project management, project financing, banking, engineering and* *construction, and contract negotiations. He has numerous published articles in the areas of project financing and administration.*

William J. Jones, *a senior research associate at the Energy Laboratory, Massachusetts Institute of Technology, specializes in alternative energy technologies. He has been a lecturer in physics at Harvard University and a staff member of the U.S. Army Corps of Engineers. Among his several publications is* Ecology and Environment *(with Richard Wilson; Academic Press, 1974).*

George Kozmetsky, *a specialist in systems analysis, organization theory, information technology, and portfolio management, is dean of the Graduate School of Business of the University of Texas-Austin. He has served on the U.S. Commission on the National Data Center, NASA's Management Advisory Panel, and several other commissions and panels. He is the author of numerous publications on information technology, technology transfer, industrial innovation, and management of large-scale systems.*

Jeanne Krause *is an historian and writer whose major interests are economics and science and technology. She was director of communications for the First Boston Corporation and is a former associate editor of* Fortune *magazine. One of her research interests has been land speculation along the Erie Canal in the early 19th century and the local impact of this macro-engineering project.*

Andrew C. Lemer, *a civil engineer and senior associate with Alan M. Voorhees & Associates (Planning Research Corporation), has worked on the physical, policy, and financial planning and evaluation of a variety of projects and programs in the United States and abroad. In addition to planning Nigeria's new capital, he has been involved in studies of major airport needs (Australia), problems of national goods movement (Iran), and financing of river crossings (Tidewater Virginia). His particular interest is in the appropriate role of large projects in development, as mechanisms of technology transfer, and in effecting institutional and social change.*

James A. McHale, *a consultant for agricultural products companies, has been a farmer for more than 30 years. He was Secretary of Agriculture in Pennsylvania from 1971 to 1976 and later became a member of the Governor's Office of Planning. His particular concern is land rebuilding and the present and future impact of macro-agricultural systems on soil quality.*

Francis Morse *is associate professor of aerospace and mechanical engineering at Boston University. His current interest is the future of the rigid airship, and he has published articles on airships and airship propulsion in journals around the world. He is a member of the American Institute of Aeronautics and Astronautics and the Lighter-than-Air Society.*

Arthur C. Parthé, Jr., *division leader for Energy Programs Technical Management Support, Charles Stark Draper Laboratory, holds advanced degrees in both mechanical engineering and business administration. He is chairman of the MIT Venture Forum, teaches courses in management information systems and management planning at Rivier College, and is co-developer of macro-engineering courses at MIT. He has authored several papers on project management and the application of modern control theory to macro-engineering projects and is a co-developer of MACRO-PLANNER--A System for Planning and Control of Macro-Engineering Projects at MIT.*

John A. Seeger, *associate professor of business administration, Northeastern University, has also been a lecturer on system, world, and urban dynamics at Northeastern University's Graduate School of Engineering. He was formerly administrative officer, System Dynamics Group, and assistant director, Division of Sponsored Research, at MIT, and has been an officer in various commercial enterprises. His current interests include simulation in policy analysis, managerial models, and organization design and the implications of the limits-to-growth debate.*

J. Peter Vajk *is a senior scientist at Science Applications, Inc., specializing in space technology assessments. He is the author of numerous articles in both technical and popular journals on the long-term implications of space industrialization and colonization on global socioeconomic development. He was a major contributor to the NASA study, "Space Industrialization: 1980 to 2010" and was a participant in the NASA-Ames Summer Study on Space Manufacturing with Non-Terrestrial Materials.*

Acknowledgments

The Editors gratefully acknowledge the initiative and counsel of Sections M (Engineering) and P (Industrial Science) of the American Association for the Advancement of Science, and of the American Institute for Aeronautics and Astronautics. As joint sponsors of the Symposium which led to this volume, they deserve the appreciation of both scholars and practitioners.

Kathryn Wolff and the staff of the AAAS Publications Office provided original suggestions on a multitude of matters of taste and detail. They were unfailingly generous with time and insights even (or especially) when under severe pressure because of a congestion of deadlines.

Beverly Bugos and Edward Marvin of the macro-engineering group at M.I.T. furnished valuable technical assistance in the preparation of portions of the manuscript.

In the launching of this new and burgeoning field of inquiry, special mention must be made of the early support of Ernst Weber and A. George Schillinger, and of the generous participation of A. Ranger Curran, Paul H. Nitze, and Ellis Scott in the first (1978) AAAS Symposium. We are delighted that Arthur Herschman and his talented staff have made arrangements for a further Macro-Engineering Symposium at the 1980 AAAS Annual Meeting to be held in San Francisco.

Cherie Wallett typed the entire manuscript, unified presentation styles of key words and phrases, and collated texts and drawings from twenty contributors with indefatigible patience, tact and good judgment.

The Brighton and Rottingdean Seashore Electric Tramroad. Rails were laid directly on the sea-bed and the passengers rode well above the waves in a wagon raised on stilts above the axles. The tramroad was described by R. St. George-Moore in "A Through-Sea Railway," Public Works, Vol. 1, July-October, 1903 (St. Brides Press, London).

Overview

There is an urgent need to develop engineer-managers who can relate technology to the deeper values of modern society. This implies broadened educational curricula and innovative career paths to provide thoroughgoing competence in engineering and management while offering more than a superficial exposure to "the other culture" of arts, letters and law. The scale of modern macro-engineering endeavors and the "external" costs associated with them have led to substantial government involvement, even in free enterprise societies. Frank P. Davidson concludes that because macro-engineering is so pervasive in its effects, it is an inherently political activity, and its practitioners must be prepared to deal sensitively with social and environmental as well as technical imperatives.

Introduction: Dilemmas of Scale and the Training of Future Engineer-Managers

How Big and Still Beautiful? Macro-Engineering Revisited has been contrived as a logical sequel to the first book in this field, Macro-Engineering and the Infrastructure of Tomorrow (1). Both volumes have been based on papers prepared for symposia at Annual Meetings of the American Association for the Advancement of Science. Last year's volume emphasized the ubiquity of large-scale infrastructure projects and the need for a better perspective on their impacts and management. The present work sets out, more aggressively, to shed multi-disciplinary light on the pre-suppositions, the semantics, the issues and the psychological, economic and political implications of humanity's awesome power to engineer the future of the planet.

The combination of engineering and organizational technology was a characteristic of early civilizations (2). What is novel in today's world is the enhanced precision and pervasiveness conferred on engineering by recent progress in the mathematical and physical sciences. Not surprisingly, the proliferation of macro-systems for power, transport and communications has been accompanied by a growing critique of the effects of industrialization on the social and physical environment. Harvey Brooks summarized this critique in a useful article, "Technology: Hope or Catastrophe?" (3), in which he reminds us that many writers "attribute our troubles not to science itself, but to its embodiment in a particular kind of technology, high technology, which is equated with large-scale, centralized systems, operated and controlled by large, impersonal bureaucracies, both corporate and governmental". He cites the paradox of plentiful food and public hygiene - both products of science and engineering - as root causes of the population explosion and its consequences of undernourishment and disease! Thus regarded, science and engineering, instead of offering solutions, have become "part of the problem".

Writing a decade ago, E.V. Krick, in An Introduction to Engineering and Engineering Design (4), concluded that "analysis of the problem areas in which engineering stands to make significant contributions reveals a special need for engineers who bring to bear on problems a broader perspective, more imagination, and a higher sense of purpose than have prevailed up to now." Jay W. Forrester, addressing the National Academy of Engineering, put it more specifically: "...our educational system is not designed to produce the special engineer-manager with the high competence needed to integrate technology into our complex modern society." (5)

In this respect, industry appears to have jumped several steps ahead of academe: "engineer-management" has already become big business and is rapidly acquiring the status of a profession. Confronted with the task of planning and supervising the construction of a new urban-industrial-port complex with a price-tag that could exceed twenty billion dollars, Bechtel Corporation was able to draw upon a pool of experienced "engineer-managers" developed within its own organization(6). Meanwhile, engineering schools have, by and large, deferred to schools of business and management to provide a viable "career path" for engineering graduates attracted by the challenge of macro-system management. While business and management students are routinely equipped with the rudiments of computer technology, the graduates of engineering schools often have little exposure to management science and tend to join corporate and governmental bureaucracies at a menial level.

In fact, the preparation of cadres capable of evaluating, planning, managing and communicating about the macro-systems of the future involves several distinct but inter-related issues. The great sin attributed to "practical" applications of science in the present century is their tendency to sub-optimize, to over-achieve narrow and limited goals at the expense of the larger systems on which life ultimately depends. It is not enough, therefore, to bridge C.P. Snow's "two cultures" (the sciences and the letters) unless both cultures share an awareness of "the big picture" and its underlying dynamics. This is a tall order; it implies a more robust approach to General Education than the token courses in physical science proposed by the literary deans of our great universities. We shall have to face an "agonizing re-appraisal" of primary and secondary education and ask ourselves, as Robert Wood has done, whether it is not timely for private endowment, on a substantial scale, to rescue public school systems from mediocrity wherever determination and leadership are present*.

Just as war has been judged "too important to be left to the generals", so engineering has become too pervasive to be left to the engineers: the correct and confident use of technology is not only in the public interest, it is essential to the prosperity, the good spirit and even the future safety of the republic.

The question of the training of engineers would be less serious if it only involved the career prospects of one species of graduate and the economic health of the small group of private enterprises committed to the management of engineering. The issue becomes one of urgent public concern because large-scale technical programs absorb increasing amounts of government investment. The executive and legislative branches of federal, state and local governments play critical roles in the decision process, and of course the public itself bears an inescapable responsibility for the election - or defeat - of candidates for office. It is essential, therefore, that a more coherent and comprehensive education be provided for "engineer-management", so that public agencies, news media, research institutes, trade unions, and business corporations may all have access to persons with the necessary "mix" of engineering, biological, legal and management studies as well as accompanying work experiences echelonned to

---

*Dr. Wood <u>et al</u> have organized a "Fund for the Boston Public Schools" on the analogy of private endowments for state-supported universities.

provide increasing scope and responsibility.

The public has a vital stake in the choice of the scale and the nature of future major technological developments. With government funding or government guarantees a frequent essential, the citizen's pocketbook is affected both by the impact on taxation and by the effect of physical facilities on the cost and quality of the delivery of various services. There is, moreover, the supervening issue of public health: in an era when megapopulations; widespread air, water and land pollution; and extensive anomie are fashionable topics, it becomes imperative to assess the indirect as well as the more easily foreseeable impacts of major investment decisions. Macro-engineering is rapidly becoming a planetary norm. What are the opportunities to employ our new, advanced technologies on behalf of individual, family and community health and well-being?

Historically, large-scale structures have had wide appeal because of their economy and the highly visible benefits resulting from such achievements as the Roman aqueducts, the Dutch dykes, and modern railway and communications systems. In certain societies there was a tradition and a mystique surrounding public works: in Rome, it was not unusual for wealthy citizens to build public baths or other facilities much as today's philanthropists endow research foundations and university chairs. In France there is still a residue of enthusiasm for "les grands travaux": Riquet de Bonrepos, the country gentleman who obtained a concession from Louis XIV to build a canal connecting the Atlantic Ocean and the Mediterranean Sea, remains something of a national hero. In Victorian England, civil engineering enjoyed a heyday, with Isambard Kingdom Brunel building the largest vessel ever launched to that time (1858) - the Great Eastern - and completing the first true underwater tunnel (under the Thames). It was, of course, a period of technological optimism. Today, it is seldom sufficient to establish technical and financial feasibility: the engineer must know in advance as much as he (or she) can about the impacts of a project on ecological factors, on people's perceptions and behavior, on energy supplies, and perhaps the international balance of payments.

In short, it is essential to understand the system dynamics of a society where engineering and politics are inextricably linked.

The risk of learning so much about so many things is that we may soon become "a society of evaluation" and lose the capacity for decision. This need not be the outcome if we build a better perspective on the central question of appropriate scale: if all large enterprises are stymied or torpedoed merely on the ground of size, we shall clearly have arrived at a pathological point in our history. Curiously, it is the late E.F. Schumacher who best understood the dilemma. Had his well-known work been provided with a different title, it could have served as a basic text in macro-engineering. In Small is Beautiful(7), he stated:

> "We always need both freedom and order. We need the freedom of lots and lots of small autonomous units and, at the same time, the orderliness of large-scale, possibly global, unity and coordination... What I wish to emphasize is the duality of the human requirement when it comes to the question of size: there is no single answer. For his different purposes, man needs many different structures, both small ones and large ones, some exclusive and some comprehensive.... What scale is appropriate? It depends on what we are trying to do."

For the engineering industry, the proliferation of gigantic projects has posed grave problems of recruitment, training and management. The Harvard Graduate School of Business Administration has sponsored a series of thoughtful papers by Professor Mel Horwitch and others on the dilemmas of "Large-Scale, Public-Private, Technological Enterprises".

> "The rise of such enterprises is a worldwide phenomenon... These endeavors possess both public and private components. They are also usually established to accomplish some mission, although such a mission can be as focused as landing on the moon or as ambiguous as implementing an overall energy program.... Society

> confronts an increasing array of large-scale, often technologically innovative needs which are beyond the technological, financial, political, and managerial capabilities of any single organization."(8)

On the subject of the huge enterprises having to do with the exploration and use of "outer space", R.C. Seamans and F.I. Ordway published a landmark report, "The Apollo Tradition: An Object Lesson for the Management of Large-Scale Technological Endeavors."(9).

The National Science Foundation, through its Science and Technology Policy Office acting "in support of the Director of NSF in his role as Presidential Science Advisor" commissioned a major study, "Constraints on Large-Scale Technological Projects", whose Summary Problem Statement is worth citing in some detail:

> "Large-scale technological projects (LSTPs), though never easy to achieve, often have enjoyed overwhelming success in the past. There is reason to believe, however, that as the industrialized societies mature it may become increasingly difficult to initiate LSTPs, or to carry them through to successful completion. Ironically, this comes at a time when the technical capacity to carry out such projects has never been greater, while the need to rely on them to deal with rising social crises has never been more urgent. Evidence for these difficulties in the form of technical failures, long delays, abandoned projects, growing regulation, and shifting public attitudes continues to mount. The consequences could be grave if major problems go unsolved..."(10)

The United States, at this moment in its history, faces a two-fold challenge. First, the national slippage in productivity and competitiveness must be made good by a concerted effort of all sectors of society. And second, we must demonstrate that our technological capacities can be directed, wisely and effectively, on behalf of human needs and aspirations. Both tasks require more than even the most rigorous analytical models, vital though these may be. We must look for such creative and adaptive energies as implied by Buckminster Fuller's familiar quest for "a comprehensive, anticipatory design science", and we must seek out the reinforcing medicines of integrity, persistence and a sense of public service. Macro-engineering is not a closet industry; there is an obvious and unavoidable political dimension.

For instance, the President and the Congress have an interesting opportunity to join with Canada and Mexico in cooperative planning for the development and conservation of the natural resources of the North American continent. Professor T.W. Kierans of Memorial University in Newfoundland has suggested one of many possible schemes for the mutually beneficial sharing of water resources: bearing in mind the dismal reception accorded the late Ralph M. Parsons when he proposed a North American Water and Power Alliance (NAWAPA) (11), it will be a test of diplomacy - and of our deeper understanding of the psychology and priorities of our neighbors - to explore such initiatives as Professor Kierans' with tact and thoroughness.

On another level, American government and industry would do well to ponder the report of the special Parliamentary Commission set up in London under Sir Montague Finniston to explore all aspects of the British engineering industry: recruitment, training, investment, labor practices, marketing, etc. Our own country has not yet set in motion a comparable response to the erosion, frequently noted (12), of our trading position. To be sure, we have problems peculiar to our own history and traditions such as the confrontation legacy of the 19th Century anti-trust response (to practices which, at the time, merited stern resistance and regulation). Today it seems essential to design a new and more cooperative relationship among government, industry, labor and the universities. Necessarily, this broad re-grouping will call for initiatives that are inter-disciplinary, inter-professional and inter-sectoral. As the late Hubert H. Humphrey's foreign policy coordinator pointed out in the concluding chapter of Macro-Engineering and the Infrastructure of Tomorrow (13), macro-engineering can now make a substantial and positive contribution to both the domestic and the foreign policy of the United States.

I know of no better illustration of the symbiosis of large-scale concepts and organization with small-scale design and concern for the values of the individual and the family, than Peter Land's description, in Chapter Three of the earlier volume on Macro-Engineering, of "The United Nations Model Neighborhood in Peru". All over the world, the skyscraper-type of impersonal housing project has spawned disappointment and disruption. Peter Land has pointed the way, for developed and developing countries alike, to build low-cost individual clustered housing, attractively designed, as a viable and proven alternative to the monolithic "apartment house."

Such an approach, consonant with the latest findings of the behavioral sciences, may lead us to a better appreciation of the psychological importance of pleasant facilities - both urban and suburban - for active exercise and recreation. Percy Corbett's redesign of a Health Center proposed for New York City after World War II deserves careful study, as the engineering industry seeks ways to dramatize its potential for the service of a healthier and happier human future.

The 1979 Symposium in Houston, Texas, carrying the same title as the present volume, was fortunate to have had as its general chairman Dr. John E. Fobes, for many years Deputy Director-General of UNESCO. Dr. Fobes, perhaps more than any other individual, fostered international commitment to the re-forestation of denuded mountain regions. In countries such as Nepal, alternative fuel sources can surely be developed for mountain people, hitherto dependent on dwindling supplies of firewood, but whose patrimony includes several of the greatest untapped sources of hydro-electric power in the world. Here is an exemplary opportunity for creative statesmanship, for cooperative international efforts to reduce the devastating and recurrent floods which threaten the Gangetic plain. Macro-engineering, in this context, can make a deliberate and considered contribution to diplomacy and can assist in the improved "north-south" relationships and the more just international economic order which is so widely desired (14).

Responsible macro-engineering implies a comprehension of macro-economic factors, of the second-order effects of investments which are, both financially and politically, preclusive in character. The well-trained macro-engineer of the future will have a professional awareness of the leading methods of analysis, including system dynamics (15), decision support (16), and the new management techniques emerging from the broadening culture of schools of business administration (17).

Can there be, in the United States, a "career path" for managers and planners of macro-engineering ventures? Do we need a new combination of academic training and increasingly responsible practical experience along the lines of the centuries-old "ponts et chausées" system in France? Mere theoretical preparation, no matter how comprehensive, can only produce bright staff officers. Senior advisers and managers must be ripened on the tree of relevant experience. In a pluralistic society, this means that future leaders of engineering should have phased exposure to public sector, private sector and third sector responsibilities.

As for the academic arena, some useful combination of management, law, biology and engineering would seem to provide the best launching pad. A combined degree program covering undergraduate and graduate studies would be one approach worth looking at. Taking a leaf from the "cooperative agreement" long established as a mode of interagency cooperation in government, it should not be beyond the wit of democratic statesmanship to devise an agreed career path that will provide useful experiences with (as one possible grouping) the United States Army Corps of Engineers, a private engineer-management firm, a leading investment bank, a consortium of major building contractors and a powerful trade union. And as a counterweight to this purposive involvement with vast affairs, the candidate might well be required to experience the chores of a self-sufficient farm and of a small or intermediate industrial establishment!

Macro-engineering can best serve as a creative and adaptable instrument for achieving social and commercial goals if its practitioners are individuals of high culture, deep learning and broad experience. Sensitively tuned to personal and community needs, with projects and programs selected by a sustainable consensus, macro-engineering may help us re-discover the sense of a coherent national and international mission.

The present volume is indicative of the wide-ranging issues with which engineer-managers will have to contend as an increasingly interconnected world society approaches the year 2000. Topics and viewpoints have been deliberately chosen to illustrate diversity: the "real world" is incurably eclectic. However, in dividing the text into two Parts ("Part I - Evaluating the Impacts" and "Part II - Cases in Point"), the editors have tried to suggest the need for both routes--the cognitive and the experiential--in the exploration of socio-technological complexity. On the one hand, attention must be given to models, conclusions, laws and systems. In addition, there is the indispensable corrective of specific cases, of life that has been lived.

Chapter 1 discusses in a general way the financial and contractual pre-requisites of the gigantic mixed-economy enterprises so characteristic of our age. Cordell Hull, who spoke on this subject to the participants in the Houston Symposium, brings to the table his practical experience as a negotiator of macro-engineering ventures and the special *optique* of the Treasurer of one of the world's principal engineer-management firms. In the following chapter, George Kozmetsky recommends a more meticulous vocabulary in our discussions of projects, programs and systems, and modulates the lessons of his years of industrial experience with the reflections resulting from sustained service as Dean of a major school of business management. William J. Jones approaches the task of evaluation from a different perspective: that of a senior research advisor concerned with the impact--or the lack of impact--of grand designs on the persistent problems of the deprived and the disadvantaged. C. Lawrence Meador and Arthur C. Parthé, Jr. conclude the opening section of Part I with an exercise that employs the full panoply of systems analysis and computer modelling: developing the concept of a "macro-engineering state space", they proceed to demonstrate a method of graphically representing the multi-faceted sectors impacted by large-scale engineering.

In a section entitled "Values, Psychology and the Decision Process", Ramón Barquín leads off with a series of questions about the linkages of large-scale undertakings with human values. Cynthia Oudejans Harris and Hans R. Huessy assess macro-engineering from the special viewpoints of individual and social psychiatry. John A. Seeger, a young academician with experience as an industrial consultant in the United Kingdom, focusses on the decision process and, after summarizing the discussion to that point, urges the widest possible participation on the part of the impacted public: however financed or defined, major-engineering ventures affect the public interest.

The second part of the volume begins with Andrew C. Lemer's evocation of the proposal (with whose configuration he has been directly involved) for a new capital city in the interior of Nigeria. John P. Craven, from his vantage-point in Hawaii, tempts us with speculations on the use of maritime potentials for the enhancement of urban life. And James A. McHale, a practical farmer who became Secretary of Agriculture of the state with the most farms--and farmers--in the country, argues passionately for a return to the family farm and for the adoption of a national farm policy that will put the health of the soil and of the rural community ahead of commercial and industrial values.

The editors have given transportation a special section, not only because it accounts for at least twenty percent of the gross national (or world) product, but because throughout recorded history all manner of transportation facilities have provided a primary field for macro-engineering. Jeanne Krause's retrospective commentary on the Erie Canal reminds us of a time when, despite a dearth of trained engineers, a single engineering project helped catalyze the "take-off" of a young and developing country. Francis Morse challenges prevailing shibboleths and, despite the "curse of the Hindenburg", sets forth a closely-reasoned case for a whole new generation of super-dirigibles. Finally, J. Peter Vajk treats us to a structured look at alternative global transport systems which, at the turn of the century, might make two-hour trips between any two points on earth a commonplace event. If this is so, or even approximately so, the word "togetherness" will have more than psychological connotations.

The concluding chapters address issues of space utilization and development. Kenneth W. Billman and William P. Gilbreath with Stuart W. Bowen provide a detailed assessment of orbiting solar reflectors as a major world power source. Not blinking at total system costs that could eventually pass the trillion-dollar mark, I think this proposal makes crystal-clear the underlying links between macro-engineering and politics. Ewald Heer's sober report on the economies and advantages of robotization in outer space provides a final impetus to reflection on ends and means, on the allocation of necessarily limited resources among competing visions of the future, on the purposes and values which should predominate in a human society no longer sure that it is confined to its own biosphere.

In the midst of such quandaries, how is humanity to make sensible investment decisions? On a national basis, there are doubts as to how much technology is appropriate for defense, how to assure energy supplies without unacceptable hazards to public health, and how to bring about effective cooperation between government and industry to maintain adequate productivity and competitiveness. To resolve such issues, it would be helpful if there were ready-made, so to speak, a General Theory of Macro-Engineering. This volume may be regarded, for all its diffuseness, as a modest step in the direction of such a theory. If engineering has truly become so sensitive and central a topic that its parameters must be familiar to business managers, civil servants, politicians and journalists, then the full interdisciplinary resources of our universities and research institutes will have to be concentrated on an effort to educate the public in the underlying dynamics. We shall have to put behind us the naive assumption that there is an inescapable conflict between environmental protection and technological progress: the problem is one of defining more precisely the environmental goals we wish to attain and then designing the engineering and institutional systems that can appropriately support our environmental and social concepts. That there must be "trade-offs" along the way, few can doubt. But the calculation of cost-benefits which include "external" impacts is no longer a novelty. If some new institutions are needed, such as a re-conceptualized Civilian Conservation Corps to reinforce public efforts at soil conservation and forest development, the public will surely be ready to consider proposals once a logical and prudent framework has been devised and presented.

Congress called attention to this general problem area when it set up the Office of Technology Assessment. More recently, the President of France established, by a special decree, the Institut Auguste Comte with the principal mission of training managers to assess and guide "major programs of equipment". In the United States, it may soon be time for a further step, beyond the tentative studies of the Office of Technology Assessment and well beyond the crisis-oriented response to unanticipated dilemmas. Just as macro-economics provided a more comprehensive view of economic phenomena, so we may expect that a long-range, comparative, inter-disciplinary analysis of macro-engineering programs and systems will make an important and perhaps at times a decisive contribution to public policy.

References

(1) Westview Press, Boulder, Colorado, 1978, (AAAS Selected Symposium 23, edited by Frank P. Davidson, L.J. Giacoletto and Robert Salkeld).

(2) Eugene S. Ferguson in ibid., Chapter 1, "Historical Perspectives on Macro-Engineering Projects". Yigael Yadin, in The Art of Warfare in Biblical Lands - in the Light of Archaeological Study, McGraw-Hill, New York, 1963, pp. 32-35, describes the discovery, in 1954, of extensive Neolithic fortifications at Jericho "built some 6000 years before the time of Joshua." Yadin makes it clear that macro-engineering has been with us for at least nine thousand years.

(3) Technology in Society, Volume 1, Number 1, Spring 1979, pp. 3-17. Pergamon Press, New York.

(4) John Wiley & Sons, Inc., New York, 1969, p. 200.

(5) "Engineering Education and Engineering Practice", in Engineering for the Benefit of Mankind, pp. 129-144, National Academy of Engineering, Washington, D.C., 1970.

(6) Richard P. Godwin, "Planning and Conduct of Macro-Engineering Projects", Chapter 4 in Macro-Engineering and the Infrastructure of Tomorrow, supra.

(7) Harper Torchbook, New York, 1973, pp. 61-62.

(8) "Uncontrolled Growth and Unfocused Growth: Unsuccessful Life Cycles of Large-Scale, Public-Private, Technological Enterprises" with special reference to the United States SST Program and the United States Attempt to Develop Synthetic Fuels from Coal, presented to Symposium on the Management of Science and Technology, held in Rio de Janeiro, June, 1978, sponsored by COPPEAD of the Federal University of Rio de Janeiro and by IA/FEA of the University of Sao Paulo. The quotations in the text are from pages 2 and 3 of this paper.

(9) Interdisciplinary Science Reviews, Volume 2, Number 4, Heyden & Son Ltd., London, 1977, pp. 270-305.

(10) From III-1, Assessment of Future National and International Problem Areas, Volume II, February 1977, (Reprinted September 1978), by Norman B. McEachron and Peter J. Teige with contributions by Harold A. Linstone, prepared for National Science Foundation, Contract NSF/STP76-02573, by SRI International, 333 Ravenswood Avenue, Menlo Park, California 94025.

(11) The Ralph M. Parsons Company, Los Angeles, California, 1966.

(12) See for instance The Future of Science and Technology, Victor J. Danilov, editor, Museum of Science and Industry, Chicago, Illinois, 1975.

(13) David Fromkin, Chapter 14, "Macro-Engineering as an Element of U.S. Domestic and Foreign Policy."

(14) Laszlo, Baker, Eisenberg and Raman, The Objectives of the New International Economic Order, published for UNITAR by the Pergamon Press, New York, 1978.

(15) See Jay W. Forrester, "Changing Economic Patterns", Technology Review, Cambridge, Mass., August/September 1978.

(16) Peter W. Keen and Michael S. Scott Morton, Decision Support Systems, Addison-Wesley, Reading, Mass., 1978.

(17) See for instance William H. Gruber and John S. Niles, The New Management, McGraw-Hill, New York, 1976.

Part 1

Evaluating the Impacts

Section A

Finance, Management and Control

Ferdinand de Lesseps, who in 1858 founded the company responsible for building the Suez Canal. The Canal reduces the sea distance between Western Europe and India by nearly 5,000 miles. (Reprinted from Vanity Fair, November 27, 1869.)

Overview

Macro-engineering projects are becoming larger, costlier and more complex; are subject to increasing risks and uncertainties because of worldwide economic and social dynamics and frequently require the capital support of national governments and/or international organizations. Although many macro-engineering programs are being cancelled, postponed or refused, Cordell W. Hull points out that the demand has never been so evident for projects which can supply the resources and infra-structures of the future. New techniques are emerging to design the financial sponsorship of large-scale technological ventures so that risks are widely and equitably shared. Much effort must be devoted to the analysis of contingencies and impacts in order to satisfy the sophisticated exigencies of the providers of "macro-finance."

Cordell W. Hull

1. Macro-Projects and Macro-Finance

Introduction

In the past few years, there has been an immense growth in the size and complexity of major industrial, energy, infrastructure and natural resource projects. Increasing costs, growing economies of scale, declining capital productivity, worldwide inflation, less economically attainable resources, higher developing-country indebtedness, remote locations, new technologies, supporting infrastructure requirements, expanding government regulations and environmental concerns are serving to make the large-size project increasingly difficult to develop and finance.

The undertaking of such macro-projects requires massive resources and involves numerous parties, all of which must be closely coordinated in the development, execution and operational phases of each such project. This includes the marshalling of many resources and integrating into the project, on appropriate terms, equipment and raw material suppliers; investor-owners; government export credit agencies; international, commercial and development banking institutions; shippers of equipment, raw material and final products; engineering/constructors; and purchasers of the project's end products. Increasingly, the host-government for the project is involved as an owner, or a guarantor of project debt or, at the very least, in a multitude of regulatory aspects; often it is the owner of the natural resources connected therewith. Frequently these macro-projects cut across national boundaries, requiring international assistance and cooperation from disparate sources.

The period from gestation of the project concept through planning, engineering and construction to operational start-up generally extends many years and often is a decade or more in today's world. Consequently, development and funding of these projects require consideration of the interests and concerns of the diverse participants, forecasting of changing economic, technological, market and resource implications over a prolonged time frame, and an appreciation of the impact of powerful geopolitical, social and environmental forces.

In order to undertake such a macro-project, all aspects must be carefully assessed and planned in advance. This involves extensive feasibility evaluation, obtaining commitments of natural resources and marketing outlets, engineering studies, and regulatory compliances. Project partners must be located, funding commitments sought, and negotiations undertaken with the concerned government agencies. Development and carrying costs to time of start-up will be very high and the various costs of delay can quickly impinge upon the project's economic viability.

This paper examines some of the major facets of funding and development that impact upon macro-projects in the context of today's international environment. These include economic, technical, financial, governmental, and management concerns. The planning and provision for these concerns determine the ultimate feasibility and success of the macro-project.

Current Economic Issues

Increasing costs, worldwide inflation, the growing scarcity of energy and natural resources, government regulations, the pace of technological innovation, currency instability, capital funding and allocation problems, and many other factors have combined to create a complex economic environment in which to undertake a project.

During the ten-year or so long period of development required, about which I spoke previously, what will be the inflation rate impact on costs of labor and materials for developing the project--five to ten percent or more per year--for materials and services purchased in the industrialized countries? Make an estimate--there can be an error of 100 percent or more in project cost if the estimate is wrong. Further, price inflation in the developing host country where local material and labor may have to be obtained can range from 25% to 75% and up per year. These rates result in a dramatic compounding impact on project costs over the period of years we are considering. It is probably reasonable to assume that inflation will continue to increase capital project costs materially over the next ten years. Furthermore, the impact of interest charges on loan funds used for development and construction rapidly mounts over the extended period. For the macro-project, the interest during construction and the escalation costs combined will often equal or exceed the initial estimated basic project cost. In addition to the normal prolonged period of development which may be expected with the macro-project, further delays can sometimes result from management inefficiencies, financing difficulties, engineering, procurement and construction problems, changes in marketing conditions, governmental intervention or regulations, and force majeure occurences.

The nature of costs, engendered by delays, depends upon the stage of development (planning, engineering, construction or start-up) affected by the delay. Delays occurring at the outset of a typical project impact costs primarily because of the escalation in the price of materials, labor and other requirements--such as insurance and taxes. As project delays occur during the latter stages of development, the associated costs are more a result of accruing direct capital costs such as interest and financing fees.

A major indirect cost is the loss of return on invested equity resulting from the delay in operations and the foregone generation of revenues and income. Where a finite resource is being developed over, say, a 10-30 year period, this return may not be lost but rather delayed. However, after giving effect to the time value of this return, its discounted or present value may be greatly diminished.

There is another type of indirect cost that is important but often overlooked. Certain project payments such as property taxes, income taxes or other taxes, various types of royalties and license fees, and other cash flows to the host government will be delayed or lost if the project is slowed. This is particularly important where a major resource development is involved. The total amount of all such added delay costs can exceed $200 million for each year of delay for a $1 billion macro-project. So, lost time is devastating to project economics!

Examples of rapidly escalating costs which serve to decrease real production capacity per construction dollar unit cost in macro-projects abound.

The Asahan hydro-power and aluminum project in Sumatra, Indonesia, being sponsored by Japanese-Indonesian interests, has increased in cost almost five-fold since 1972 ($420 million to over $2 billion).

A coal gasification project of 250,000 c.f. per day capacity costs $1 billion today, over double from 1973 and the gas production cost connected therewith is now estimated at $3.00 per 1000 c.f. compared with $1.32 when it was in the planning stages.

It costs a billion dollars to open a 150,000 ton per year copper mine, in a developing country, including infrastructure--triple its former cost in 1973.

So, in the face of the various uncertainties, capital hesitates. It turns to less indeterminate outlets and seeks quicker returns.

Yet supply demands, which the macro-project will need to service, continue to mount. Worldwide, in the mining area, much new capacity will be needed between now and 1990 for the development of iron ore, bauxite, alumina, nickel and other metals.

Growth in the demand for many such natural resources could increase four to five times through the end of the century. Worldwide, the energy industry will require many billions of dollars for oil and gas developments, nuclear plants, coal projects and the new emerging technologies in the solar, geothermal and synthetic fuel areas.

Will these projects be developed when needed? If not, the impact upon the world economy will be substantial. Higher inflation and lower GNP growth and higher unemployment will be the inevitable result.

Turning to support for these projects from various areas of the world, we discern that European countries are reducing direct investment in macro-projects in both Europe and the developing world due to a chronic slowdown in economic growth. However, this adverse trend is perhaps more than offset by increasing amounts of debt financing made available to such projects on preferential terms through the medium of export credits supporting the sale of material and services for the development of such projects.

In the United States, macro-projects in almost every sector are continuously being delayed because of:

- Economic uncertainty
- Capital unavailability
- Lack of an energy program
- Uncertainty on future energy and natural resource demand and prices
- Increased government regulations and environmental controls
- Questions on nuclear energy
- Unresponsive Public Utility Commission attitudes concerning appropriate rate of returns, allowances for work in progress in rate bases, etc. for new major power projects
- Inadequate programs for export support

Further, because of various uncertainties including the changing nature and philosophies of developing countries toward ownership and development of their natural resources, the flow of investment capital from the United States for the development of macro-projects in the Third World has faltered greatly in recent years. However, it does appear that debt funding for such projects insofar as they use U.S. goods and services is now becoming more available from the U.S. Export-Import Bank.

In the last decade, Japan's rapid economic growth supported the development of many major energy and natural resource projects in Third World countries. While direct investments in such projects have dropped drastically, Japan's aggressive export program, supported by advantageous export credits, is still providing very considerable funding impetus to the development of international macro-projects. Also, Japan is currently considering increasing its foreign investment program.

All of these types of factors affect the project's economics and funding availability and emphasize the need for detailed master and feasibility planning, and in-depth economic studies of the project, the markets and the forward longer term economic environment into which the project must fit. Increasingly, efforts must be made to reduce project delays and, as never before, the most professional talent available for macro-project development and implementation must be brought to bear throughout the feasibility, planning, financing, engineering, construction and operational phases.

Capital and Funding

Where can the enormous sums required for funding the macro-projects be located?

In the recent past, projects could be funded, to a large degree, directly by the sponsors from a mix of capital reserves and current cash flow. These funds were supplemented by debt through banks or institutional lenders, usually guaranteed by the sponsoring corporation. However, due to the increase in scope, cost, risk, and number of projects, owners are unable to follow such a traditional financial approach, certainly with respect to the macro-project. Instead, much of the required funding support for such projects is in the form of debt from traditional capital markets, eurodollar sources, or governments. Further, governments aid project development by guaranteeing loans and by allocating funds in the form of grants, concessional lending, export credits and investment guarantees. A macro-project is generally so capital intensive that it must be financed from a variety of such sources, including international commercial banks, eurodollar lenders, local currency sources, institutional lenders in the United States or Europe, export credit agencies, the World Bank, development banks and the like.

Putting together an appropriate funding package from such diverse institutions requires an intimate knowledge of the world's financial markets, an appreciation of lenders' concerns, requirements and preferences, as well as creative and responsive techniques in structuring the elements of a project to qualify for the type of loan which meets with the project's amortization ability.

As such projects increase in size, both in absolute terms and in relation to the capital of potential corporate participants, the trend toward consortia and joint ventures of multi-sponsors in developing and investing in such projects may be expected to continue. For example, this may take the form of a $500 million copper project such as Bouganville where a multinational group of companies and banks from the U.S., Canada, the U.K., Japan and Australia joined together to develop and finance the production of 160,000 tons per year of copper contained in concentrate; and it may also take the form of a $3 billion integrated steel mill, such as Aconimas in Brazil where a multinational group of companies from the U.S., Europe and Brazil joined together to develop an installation to produce two million tons of ingots per year, plus 1.5 million tons of products, and it may also include direct ownership participation by the host government, as also is the case with Acominas.

This approach has many advantages. First, it serves to broaden the sources of equity and increase the amounts available for this purpose. Second, it spreads the risks of project completion and operation, and provides more sources of additional funds to cover any project cost overruns. Third, it would seem to lessen the risk of arbitrary actions, such as expropriation by host governments, since such actions would impact nationals from various countries, all of which ostensibly may be affronted by such actions. Finally, it provides a broader base of development and operational expertise, diversifies the marketing base and outlets for the project's product and increases the sources of export credit assistance and bank funding for development of the project.

As to the sources of loans, they continue to proliferate. Liquidity in the eurodollar market is being exponentially expanded by such factors as the United States' balance of payments deficits, recycling of Arab petrodollars, the multiplier effect of fewer reserve or capital requirements of banks operating in this market, and continuous repayments of outstanding eurodollar loans. Many of these eurodollars are available for loans or investments in macro-projects.

The balance of payment problems of the United States and stagnated local economic growth in Europe and Japan have increased amounts of export credits available for macro-projects to take advantage of the continued absorptive power of international projects for exported goods and services. New sources of funds are opening up such as from the rapidly evolving financial markets of the Arab world as well as in the form of government credits supporting the export of services and materials from developing countries, such as South Korea, Brazil and India. Also, since the debt of developing countries is high (over $200 billion), commercial banks are increasingly directing their attention to feasible foreign exchange generating projects whose integral commercial viability and internal cash flow can support additional loans, permitting direct or indirect funding for activities within developing countries already heavily indebted to continue selectively. Furthermore, through participating in such projects and becoming involved in financial planning, letters of credit, fund transfers, performance bonds, interim contractor financing and the like, banks can greatly increase their own returns over instances where they simply participate as lenders in thinly margined balance of payment loans.

Major equipment, raw material or prospective market product offtakers from the project are increasingly willing to make loans or advance payments on projects to secure markets or sell their products. The World Bank is increasing its lending in the energy sector. So far the institution has made over 300 loans totalling over $9 billion for power projects and by 1990 such loans are expected to reach upwards to $1 billion a year. Since the Bank only lends a portion of the project's costs, this could indirectly generate additional cofinancing from other sources exceeding $3 billion a year. Loan packages which involve parallel or joint arrangements with the World Bank also provide the advantageous involvement of the World Bank's highly qualified staff in assessing the project and helping monitor loan drawdown and project progress.

Funding Source Concerns

Meeting the requirements of the financial institutions requires a financially sound credit and a well conceived project, with provisions for timely cash flow generation, adequate returns for investors, appropriate debt coverage ratios and adequate security for the debt. Each funding institution connected with the project will demand a financially sound credit.

Some of these sources only provide medium-term loans while others, such as bond markets and export credit agencies, provide longer term funds. The various sources of funds must be appropriately blended so that the project cash flow, currency of repayment, loan repayment tenures, form of security and interest rate structure satisfy the requirements of the project. This presents an extremely difficult task in connection with the macro-projects for several reasons.

First, there is a question as to loan security. Rarely are the sponsors of the macro-project of sufficient size that they could absorb the full impact of the debt on their balance sheets. So, some form of off-balance-sheet project financing is usually necessary and the lenders' prime security increasingly depends on the cash flow and commercial viability of the project instead of the more diversified and broader resources of an already established business. This immediately raises the question as to how risks and benefits should be allocated among the owners of the project, the host government, the market offtakers, the supporting engineer/contractors, and the other involved parties as well as how their contractual and economic relationships should interrelate. In assessing the credit and the probability of loan repayment, the lenders must be acquainted with the raw material arrangements, economics,capital costs, marketing arrangements, and cash flow of the project and be satisfied that they are adequately and appropriately covered for cost overruns which may be incurred in completing the project, late completion, failure to complete, post completion shut-down, governmental interventions, expropriations or complete project failure. The macro-project sponsors must present a complete and detailed evaluation of each of these risks, and outline precisely how they are to be covered. Generally speaking, this will often result in the project sponsors' unconditionally guaranteeing completion of the project and committing for any additional funds which may be required for that purpose. In addition, they may have to agree to fund working capital or funding deficiencies for stated purposes up to agreed upon amounts during the actual operation of the project itself. Guarantees from the host government are frequently sought or obtained for a portion of or all of these types of risks, either because it has a vested direct interest in the project or because economic benefits in the form of taxes, duties, production of foreign exchange, development of resources, and training and employment of nationals will result from the project.

Regardless of such interim guarantees and assurances, it is fair to anticipate that once the project has commenced production operations, has proven its viability and sound marketing commitments have been arranged, that certain of the initial security undertakings may be allowed to fall away and lenders may be more willing to look to the security of the project itself including the cash flow being generated. This is particularly so if feedstock is assured at a given pricing formula, operating costs have been established and long term contractual arrangements are in place insuring marketing of the product at profitable margins during the period of debt repayment.

Lenders are rapidly increasing their capability to understand and perceive project risks. Accordingly, they are rightly insisting on more detailed feasibility, economic, and engineering studies of all the risk variables. The more thoroughly these elements are presented and sensitivity factors developed, showing the effects of changes in costs of construction, revenues, interest rates and operating costs on returns on equity and ability to repay debt, and appropriate arrangements demonstrated for the involvement of qualified, experienced firms in studying, developing and operating the project, the more readily an appropriate funding package will fall into place. Increasingly, I predict that lenders will insist on higher compensation for their money and services as they inevitably assume some of the risks themselves which are a part and parcel of major projects.

A most important factor in project evaluation is the market projection for the project product when it starts up, ten years or so out. What will the world economy look like then? Will there be a prolonged period of slow economic growth? Will market conditions have changed? Will there be product substitutions or technological breakthroughs rendering the product obsolete or changing the project's economics? Will there still be a hospitable host government?

For the macro-project these are exceedingly difficult but imperative questions. Because of the long period for developing these projects and the uncertainties in the economic environment, lenders have become increasingly wary about the possibility of cost increases, delays in completion and the ability of the project to market its products far in the future to produce adequate revenues, leaving appropriate margins of safety and cash flow available for debt service. Accordingly, increasing attention must be given to working with lenders so that each of these risk areas is thoroughly studied and understood by all the parties concerned with the project, that all reasonable risks have been considered and allocated among the parties and elements of the project, and adequate assurances have been provided that either the risk is remote or that it is adequately secured against.

In today's macro-projects, an adequate rate of return must be allowed in the initial projections to provide a reasonable residual rate of return to the project's sponsors, provide a satisfactory margin of coverage for the debt and allow for various contingencies which the macro-project will inevitably encounter, considering the indefinite economic situation in the future when the macro-project will come on stream, the immense size of the project and the very considerable time including both anticipated and unanticipated delays, which will occur prior to completion. Also, the unknown applicable inflation rates and other cost factors of the project, the currency exchange risks, and potential local host country government risk because of indefinite and unknown prospective policies within the host country bear directly on the required returns. Whether macro-projects are really being structured to provide appropriate returns, taking into account all of the various risk elements is indeed questionable and is another area of increasing concern to the lenders and to the investors seeking positions in such projects.

Turning to the currency exchange risk question, because of the necessity to blend together so many sources of funds for the project, an international macro-project often ends up with debt commitments in at least several and often many different currencies. While it is impossible to project with any degree of certainty over such a long time, the cumulative impact of currency movements on the project, it can be taken for granted that currencies will move with respect to one another over the time frame in question and that such movements will have a most substantial impact on the economics and debt servicing capabilities of the project. Without attempting to forecast precisely what this impact might be, it is good practice to analyze various possible ranges of currency movements with respect to the currencies in question and determine how, at least over the period the debt is outstanding, they will impact on the project's economics. Often adverse movements will be shown to erode substantially into viability and may militate against committing for debt in certain currencies, or result in a preference of one source of loan over another. Ideally, the currencies of debt should be matched as nearly as possible to the currencies of revenues which will inure to the project during operation. Since this is only infrequently possible, at least attempts should be made to maximize this feature while committing the remainder of debt in those currencies whose movement might be expected to move or remain more closely in line with the movement of the currency of revenues of the project--a complicated, frustrating and sometimes non-productive analytical task, but one that really should be done.

Other Issues

There are many other issues which I can only briefly touch upon here. With respect to macro-projects in developing countries, most governments are becoming increasingly involved either as a major co-sponsor or as sole owners, or at least guarantors of the debt, particularly in the cases of natural resource development projects.

Here, questions arise such as where will the expertise be obtained for evaluating, planning, managing, constructing and operating the project. Often personnel experienced in these types of endeavors are in short supply in developing countries. Where will the product be marketed since the country may not be effectively integrated into an effective distribution system?

The country may have broader considerations transcending project economics, i.e., national security of supply, employment, prestige, social goals, import substitutions, a wasting natural resource (flaring of gas), etc., which require the implementation of the project in spite of what may be marginal or questionable traditional project economics. How should such a project be structured and presented in competing for international capital funds?

As the periods of time for project implementation are prolonged, how is equity capital to be directed to such projects in view of quicker payout project opportunities available? The returns must be sufficient to attract such capital, and tax and other investment incentives should be considered by all concerned governments to enhance the project's viability.

What are the future implications to the country and to the world economies arising from the failure of governments to provide a hospitable investment environment, of governments failing to appreciate the vast risks inherent in such projects, failing to allow the returns necessary to attract capital to the macro-project, or failing to provide long-term regulatory certainty? These questions are directed as much to the industrialized world, in my case to the U.S., as to developing countries.

Conclusions

So, drawing several observations out of this discussion in relationship to development and funding of macro projects, I would submit for consideration the following:

1. Macro-projects are getting bigger, more complex and costly, and take increasingly longer development time.

2. The number of parties and diversity of nationalities connected with a macro-project is growing.

3. Governments are playing an increasing role in these projects.

4. They cannot be viewed in isolation and they are increasingly tied to world economic and geopolitical forces.

5. The economic uncertainty of the world affects the timing of and type of which macro-project proceeds.

6. Because of economic uncertainties, these types of projects are being increasingly deferred.

7. The cost of delays, which are most substantial, can amount to as much as 20% or more per year of the initial estimated capital cost.

8. Delays in the construction/development of macro-projects in the energy, natural resource and other essential sectors so necessary to future world economy will inevitably mean raw material and energy shortages sometime in the near future.

9. The result of this impact will be inflationary and depressing to economies and employment.

10. When developing a macro-project, increasing attention should be given to the feasibility

analysis, including sensitivity analysis of variations in the essential parameters of capital cost, revenues, cost of money, operating cost and the like, and professional studies must be made of the impact upon the project's economics and the ability to repay debt of:

- fluctuations in currency exchange rates of the various currencies in which the debt is committed and the currencies and revenue of the project;
- cost of delays;
- prospective markets;
- the international economic environment;
- the program for project implementation.

11. Increasingly, funds are sought for project development from other than internally generated or guaranteed sources. Therefore, projects must be structured so as to qualify for loans from numerous sources on a project financing basis.

12. While the sources of capital for macro-projects are increasing and becoming more diverse, capital shuns uncertainty and will be allocated where the risk-return aspects are most favorable.

13. Lenders increasingly are concerned with the risk factors such as cost of completion, delays, cost overruns, viability of the marketing arrangements, technical appropriateness of the project, and the manner in which the project will be managed during the construction and operational periods. Funds will flow only to those macro-projects which are conceived to take into account the issues which have been discussed in this paper and which have made a sound and responsive evaluation of the various risks connected with the project, have provided appropriate answers and security mechanisms with respect to those risks and have arranged for qualified and experienced personnel to study and manage the development, engineering, construction and operation of the project.

14. New techniques of structuring sponsorship of projects must be sought including novel arrangements amongst the equipment suppliers, engineer/contractors, owners of the natural resources, traditional sponsors, host governments, and market offtakers to expand the equity funding base and

provide a complete integral package of all skills and services needed to make the project a success and spread the risks while assuring each such participant an adequate return.

Note

This paper was also delivered at the Asian International Chemical and Process Engineering and Contracting Show in Singapore, January 16, 1979.

Overview

It is appropriate to develop new tools and techniques for the evaluation and management of macro-engineering efforts because the macro-systems are no longer self-contained but are highly inter-related. George A. Kozmetsky argues that traditional conceptual models for management are limited insofar as they are restricted to single objectives of an economic nature, to self-contained operational units, and to static environments. Impetus for the development of macro-systems management can logically come from the senior executives and administrators of our major institutions: these are the individuals who bear the operational responsibility for assessing policies, programs and plans. These executives have an opportunity to develop the new instrumentalities for inter-sectoral policy analysis and for the management strategies needed to maintain a viable society.

George Kozmetsky

2. Evaluation of Macro-Systems: Models and Case Analysis

The occasion of the January 1979 American Association for the Advancement of Science meeting is an appropriate setting to revisit macro-engineering and to evaluate its future as to "How Big and Still Beautiful." One of my early findings in revisiting the literature on macro-engineering is that as an epistemology, it is still to be organized and conformed. The best literature available is that to be found in the 1978 AAAS proceedings of the Macro-Engineering Symposium. Therefore, the first part of my talk this morning is to use three of last year's papers as a benchmark to establish the parameters for evaluating macro-engineering systems and then devote the remainder of this paper to selected topics and cases that I believe require development in order to mean big _is_ beautiful as well as to provide means for developing the required epistemology.

Basic Evaluative Framework

Before proceeding with the main position of this paper, it is appropriate to set forth the background I have used in terms of the general state of modeling and analysis for macro-engineering. The general area of socio-economic policy systems modeling and analysis since 1970 is proving to be one of the major areas of academic research. There is much research work being devoted by many academicians to bring systems analysis and other methodologies to bear on a solution of macro- and micro-social problems. The impact of this work on common problems may well provide the tools and means for assessing whether either or both small and large macro-engineering is beautiful and how both large and small can benefit mankind.

Professor Kenneth E. Boulding succinctly stated the need and approach to the evaluation of large systems as follows:

> "The evaluation of large systems itself is a difficult problem, one which cannot easily be resolved by simple discussion or argument. Nevertheless, these remain questions of enormous importance, pregnant with potential for vast human misery, or perhaps great human satisfactions.
>
> "Improving our methods of evaluating these large systems, therefore, seems to be a high priority for the human race if we are to avoid what might be catastrophic error and ruinous decisions. At this stage I doubt very much if any general rules can be formulated to assist in these evaluations. Nevertheless, if the description of large systems can be improved, and if we can estimate the relative evaluative weights given different parts of the system, then we have at least some chance of breaking down the problem into smaller and perhaps more manageable pieces."[1]

It has been the primary assumption in this paper that the management of macro-engineering problems provides an adequate framework to describe macro-engineering systems that encompass their assessment in economic, social, political, and technological terms. In this respect, it is my opinion that macro-engineering has progressed from macro-engineering projects, to macro-engineering programs, and that our recent concerns with environment, resources, scarcity, social welfare, and national policies related to energy and technology are evidence of this project to program shift. My precept is that throughout time such needs and issues have had to be concerns of the senior executives and administrators of our major institutions.

---

[1] Kenneth E. Boulding, "The Evaluation of Large Systems," Technology Review, May 1975, p.68.

Let me now be more specific. Managers of institutions exert authority through planning, operations, and control systems. Those systems simultaneously conform to and mold the manager's conception of institutions, their purposes and their needs.

A simplified model of a planning, operations and control system can be shown as:

Institutional Planning and Control

Resources	Planning	Operations	Control	Goods Services
	Desired Objective	Actual Transformation of Resources	Difference between plans and actuals	

In this model, inputs are defined as the resources required to implement plans. These resources are then transformed through operations and controlled to meet the perceived objectives of the institution.

Even in its detailed, real-world form the foregoing model makes it hard for managers to perceive and analyze broad cultural, social and technical patterns in terms of their institutions' desired objectives. Furthermore, the model's field of view is too narrow to aid managers in seeing their own and their institutions' roles in the context of society and its goals. In short, traditional conceptual models for management, though useful in the past, are rapidly becoming limited insofar as they are restricted to single objectives (primarily economic in nature), to self-contained operational units, and to static environments.

In other words, the traditional models are unable to provide management with the:

- means of conceptualizing and dealing with multiple objectives,

- means of conceptualizing and dealing with operations and interactions with other economic, social, cultural institutions,
- means of conceptualizing and operating within a highly dynamic environment.

Macro-engineering can give management the ability to overcome most of those restrictions. Some of the technology exists; some needs to be developed. The critical need is for constructs that will direct the advances and applications of macro-engineering so as to benefit management and the public.

It is the manner in which macro-engineering projects and programs have been managed in the past two decades that provides us with an adequate framework to evaluate macro-systems currently and for future needs.

Macro-Engineering Revisited

Three papers presented last year provide an excellent starting point to revisit macro-engineering. These papers are "Historical Perspectives on Macro-Engineering Projects" by E.S. Ferguson,[2] "The Planning and Conduct of Macro-Engineering Projects" by R.P. Godwin,[3] and "The Finance of Macro-Projects" by W.O. Sellers.[4] These papers, in some respects, established some of the more critical parameters in regard to the understanding and practices of macro-engineering projects.

Dr. Ferguson defined macro-engineering projects as follows:

> "A macro-engineering project (MEP) in any age is one that strains current capabilities and resources. It is at the outer limits of current 'state of the art'; it is expensive; and because it is so different from run-of-the-mill projects, the MEP furnishes engineers and technicians with a particular challenge and fascination."

---

[2]University of Delaware and The Hagley Museum.

[3]Vice President and Director, Bechtel Inc.

[4]Vice President, Merrill, Lynch, Pierce, Fenner & Smith, Inc.

His definition concentrates macro-engineering on large, unique projects that involve newer technology for the first time or are very expensive. Thereby, he would include projects simply because they were physically big. What is interesting to me is that Dr. Ferguson's definition could include a project such as the English Channel Tunnel as a macro-engineering project in 1751 when Nicolas Desmarets won the Academy of Amiens prize for the test plan for a "dry shod" connection with the British Isles, but not in 1979 when it is generally accepted that the tunnel could be bored by very conventional methods. It is for this reason in revisiting macro-engineering that I find R.P. Godwin's definition more useful to develop a required epistemology and practices in that he defines a macro-engineering project as "a large and complex engineering project." Godwin's examples of macro-engineering wonders are more closely allied to Dr. Ferguson's general definition of macro-engineering projects.

Godwin also distinguished between macro-engineering projects and programs. His distinctions between projects and programs included similarities and differences. Their similarities are that each is (1) complex; (2) large as to size or scope or both; (3) takes a long time to implement, generally measured in decades; (4) requires significant resources in terms of manpower, material, and equipment, resources that can be measured in the billions of dollars; (5) government involvement either as client before projects can begin and/or regulatory; (6) government involvement in financing as his lower limit for macro-engineering projects is $10 billion; and, finally (7) has substantial social, cultural, economic and environmental impacts.

The differences he sets forth between macro-engineering projects and programs can be categorized as to <u>first scope</u>. Projects are generally single-purpose facilities that are confined to a specific site. Programs are national in scope that are multiple-purpose projects that spread over several sites for the purpose of developing hardware or some ultimate product. A second category that distinguishes projects and programs is funding. Projects are funded separately and from worldwide financial institutions. Programs are funded from a national treasury. Perhaps examples of macro-engineering projects and programs can help to clarify his distinctions. Examples of projects he used are the Pyramid of Cheops, the Great Wall of China, the Panama Canal, the Jubail Industrial Complex, the Trans-Canada Pipeline and the James Bay Hydro-electric Project. His macro-engineering program examples are the Apollo Program, the Manhattan Project, and the national highway program

In my opinion, it is implicit in Mr. Godwin's differentiation between projects and programs that there is more technology that can be diffused to other societal uses through macro-engineering programs than through macro-engineering projects. Projects employ more proven state-of-the-art technology while programs are concerned with both cutting-edge technology, state-of-the-art technology and required advances of selected scientific breakthroughs. The financing, management and assessment of projects and programs can thereby be differentiated. Therefore, in revisiting macro-engineering, it is important that the epistemology does distinguish between projects and programs. Their similarities often have masked their significant characteristics.

The majority of last year's symposia was concerned more with the macro-engineering projects than with macro-engineering programs. In my preparation of this paper on the evaluation of macro-systems, it became necessary to review some case studies of macro-engineering programs for purposes of contributing towards a body of knowledge on macro-systems. For purposes of this paper, I have placed more emphasis on my review of the management, planning, and conduct of macro-engineering programs.

Macro-Engineering Programs and Institutional Complexes

Macro-engineering programs generally originated from the need of a nation to meet a single objective. More often than not, these needs arose from crises that involved national defense or national prestige and were given top national priorities. Examples of such programs were the Manhattan Project and the Apollo Program. These programs had definite end goals which could be perceived by the program managers as well as by most segments of the national public. These macro-programmatic national demands were expressed through the medium of one aggregative consumer institution-the Federal government. The government generally established a new institution to plan the program, conduct the program, and control the program including its financial, management, engineering and quality controls. In response to the social demands as exemplified by macro-engineering programs, a limited number of institutional suppliers, in terms of system managers, R & D managers, equipment suppliers, and component manufacturers,*et al.*, evolved in a relatively short period. By the time President Eisenhower retired from office, he referred to these individual supply-oriented institutions as clusters and the associated federal agencies as a complex; e.g., the military-industrial complex. He went on further to call attention that these complexes could well try to perpetuate themselves. The same general evaluative questions are being asked today by various groups of other programs such as TVA, mass transit programs, welfare, and other macro-programs.[5] In the past, macro-engineering programs have institutionalized themselves both in terms of program management and in terms of an institutional constituency comprised of individual education and business entities that view the program as a demand for their services and goods. Institutionalization of macro-engineering programs through institutional complexes is an emergent phenomenon since the 1960s.

One can model the various goals for such an institutional complex in terms of macro-systems. However, the limitations in models are encountered in the situation where we must know when to change

---

[5]"TVA Today: Former Reformers in an Era of Expensive Electricity," *Science*, Vol.194, pp.814-818. F.C. Colcord,Jr., "Institutions for Urban Transportation," *Technology Review*, Oct/Nov. 1973, pp.58-59. A. Altshuler, "The Politics of Urban Transportation Innovation," *Technology Review*, May 1977, pp.51-57. E. Herst, "Transportation Energy Conservation Policies, "*Science*, Vol.192, pp.19-20. R. Morris, "Welfare Reform 1973: The Social Services Dimension," *Science*, Vol. 181, pp. 515-522.

the goals that the macro-system represents. To date, there are no models to handle these classes of problems. Part of the reason for the lack of such models is that the current models were based on optimal planning solutions to societal problems which these programs then represented. Because macro-engineering problems take a long time for solution the emphasis has been on successful solution of the problems of a previous period and not on the impact that their solution created on present and future needs.

Willis W. Harman and his colleagues at Stanford Research Institute have identified, in their analysis of social problems, the following four dilemmas inherent in past macro-system solutions; namely[6],

1) The growth dilemma. This dilemma centers on the need for continued economic growth but that our society cannot live with consequences; e.g., environmental and social costs.
2) The control dilemma. There is a need to guide technological innovation but we shun centralized control.
3) The distribution dilemma. There is no suitable mechanism or philosophy within the industrial system for redistribution between highly industrialized nations or regions with the less developed nations or regions.
4) The work-roles dilemma. Our society is increasingly unable to supply an adequate number of meaningful social roles.

These four dilemmas make up what Harman refers to as the world macro-problem.

> "Industrial societies in general, and the U.S. in particular, are faced with one fundamental problem which is so pervasive and so pernicious that the related societal problems (e.g., poverty, unemployment, inflation, environmental deterioration, crime, alienation) will defeat all attempts at solution until it is satisfactorily

[6] W.W. Harman, "The Coming Transformation," The Futurist, February 1977, pp.5-7.

resolved. The problem is not new, but industrial and technological advances have given it a new urgency.

"Individuals, corporations, government agencies and others in the course of their activities make <u>micro</u>decisions (e.g., to buy a certain product, to employ a person for a particular task) which combine to form <u>macro</u>decisions of the overall society (e.g., a 5% growth rate, deteriorating cities, polluted air.)

"The world macroproblem now is that <u>perfectly reasonable microdecisions currently are adding up to largely unsatisfactory macrodecisions</u>. The macroproblem will be the predominant concern of the foreseeable future for all alternative paths. It is the composite of all the problems which have been brought about by a combination of large-scale industrial development and high population levels.

"Although such terms as 'environment' and 'technology assessment' have entered the political rhetoric, we have not yet begun to take the macroproblem seriously."[7]

Macro-Engineering and Social Engineering

For economy of time, let me generalize what I believe is a next evolutionary step within macro-engineering. In this regard, it is necessary to explicitly recognize the implicit hierarchy underlying this discussion. This hierarchy is macro-engineering projects, macro-engineering programs, and macro-systems. The linkage among these three levels of macro-structures is contained in the management dimension. In my view, macro-engineering management evolved from macro-engineering projects management to macro-engineering program management and is now evolving macro-systems management. The implication here is that the principles of macro-engineering project management are developed at the macro-engineering program level. Analogously, the principles of macro-engineering program management are to be developed at the macro-systems level.

---

[7] Ibid, pp. 7-8.

The management of projects is well recorded in the literature and there are sufficient good and bad examples for study. To a lesser extent this is also true of macro-engineering programs. However, there is need to identify, to evaluate, and to extend our knowledge base with regard to macro-systems management. Macro-system management is to study the manner in which government, business, and educational institutions can be utilized to meet the demands of society that involve large-scale needs such as security, environment, education, and social,cultural and natural resources while maintaining a free, modern society.

Another way of making the same point is to say that micro-engineering must be co-joined with social engineering at a macro-system level so that newer institutional arrangements -- economic, political, social--can be formed that solve today's and tomorrow's human and social needs. This will require newer ways of establishing multi-goals for macro-systems. The goal setting is no longer simply in economic terms or social terms, but must establish political, technological and cultural goals.

An example of social engineering was provided by Jack C. Page, vice president of Booz, Allen and Hamilton, Inc.

> "Air and automotive travel have inherent advantages over rail travel. The management and labor practices in the respective industries affect their performances. But when it takes longer to fly from New York to Philadelphia now than it did 30 years ago, and when our cities have monumental traffic jams, it is obvious we have not reached the optimum balance among these three modes.
>
> "Suppose that gasoline tax revenues were used to develop better railroad terminals, reservations, signalling and communications equipment; railroad taxes to finance aircraft navigational aids and landing facilities; airline tax revenues to build new roads and super highways. What would happen?

> "Under present conditions, the construction of new aircraft navigational facilities and terminals would almost cease while revenues would still flow in to complete the highway building program. Meanwhile, the railroads would have substantial funds to improve until they could compete. Their success would again provide revenues to improve the airlines. A stable transportation system would ensue; and the relative emphasis on different modes would be determined by the relative tax rates. Raising the gasoline tax would improve rail service which would, in turn, improve air service, and a new stable operating condition would exist with decreased emphasis on the automobile."[8]

Another reason for looking at macro-engineering from a macro-system and its institutional complexes is that it can help develop required national policies. J.E. Goldman, senior vice president for R & D at Xerox Corporation stated the following regarding development of a national technology policy:

> "As we look back over the past decade, we have good reason to be uncomfortable with our national investment in technology... My own reasons for concern are twofold.
>
> "First, it seems to me we have given far too little support to the creation of new technologies, including too little toward the support of science itself. Rather, we have spent our dollars--and the energies of our technical people--on the exploitation of given features of the old technologies, principally through scale-up and increasing reliability.
>
> "Second, our technical priorities have held relatively stable, despite the fact that the world about us is undergoing enormous change. During this past decade, to cite one specific fact, our national government has invested some 3 billion

---

[8]Jack C. Page, "Engineering Social Systems," Technology Review, July/August 1972, p.46.

to 4 billion man-years in research and development (R & D) programs; less than 1 percent of this enormous investment has gone toward R & D support relating to such critical problems as housing, crime, the urban environment, and ground transportation.

"As we correct this imbalance and begin now to channel a larger part of our technical effort toward programs of social significance, we must remember one of the important lessons of our recent past: the history of the space program, it seems to me, is a lesson in the mastery of the institutional techniques necessary to bring together the segments of the intellectual, industrial, and technological community needed to fulfill goals in a timely fashion."[9]

Macro-Engineering Financing and Policy Initiatives

The financing of macro-systems is as complicated as the assessment of macro-engineering projects, programs and systems. Wallace O. Sellers, in his 1978 macro-engineering symposia paper entitled, "The Financing of Macro-Projects," pointed out that there are over 200 macro-projects that were in the hard planning stages or under consideration which will cost over half a billion dollars each. Many today are over $10-25 billion each. Energy program investments will easily consume over 30% of all savings over the next decade.[10] While Sellers indicated that financial institutions were "no longer intimidated by large numbers," he did set forth the factors which needed to be considered before successful financing could be achieved. Again, these factors require

---

[9] J.E. Goldman, "Toward a National Technology Policy," Science, Vol. 177, p.1078.

[10] G. Kozmetsky and E.B. Konecci, "National Energy Plan and Investment Analyses," Preliminary Assessment of the President's National Energy Plan, May 11, 1977, The University of Texas at Austin.

inter-institutional macro-system coordination, planning, construction, and control.

Today there is urgent need to consider and develop integrated macro-system management. The burden of responsibility and the authority to initiate the needed macro-systems rest squarely on our senior executives. They not only have the formal ability to reallocate resources but are, in fact, the single key element required to guide and to assess the policies and programs and plans in a rational manner.

The challenges of the macro-system programs will not be met or overcome by the use of polls, holding symposia, or conferences, conducting panels of expert predictions or in the use of casual "business as usual" staff work. Nor will they be solved by external assessment expressions such as holding protests, lengthy court battles, etc. The urgency of the problems themselves will no longer permit such luxury in the management of change. The methods must change. The requirement is for personal involvement at the levels of senior executives who are concerned with their own institutional policy formulation and decision making. Selected groups of executives from the major institutions must work in collaboration and in concert to formulate the programs and to take the first steps to resolve them. These working committees of executives are not one-shot affairs. They are continuing affairs. We know other nations have developed policy centers as working committees of senior executives; in particular, Japan. Japan has established the following policy centers:[11]

1. Public and Financial Subcommittee
   a. Working Group on Allocation of Investment
   b. Working Group on Social Overhead Capital

2. Industry Subcommittee
   a. Working Group on Information
   b. Working Group on Resources

[11]Taken from "New Economic and Social Development Plan 1970-1975," Economic Planning Agency, Government of Japan, pp. 163-185.

c. Working Group on Progress of Technology
d. Working Group on Labor Force
e. Working Group on Small Business and Distribution
f. Working Group on Agriculture

3. National Livelihood of People Subcommittee
   a. Working Group on Level of National Livelihood of People
   b. Working Group on Environmental Pollution
   c. Working Group on Urbanization
   d. Working Group on Social Security
   e. Working Group on Social Tension

4. International Economy Subcommittee
   a. Working Group on International Capital Movement
   b. Working Group on Economic Cooperation

5. Research Committee re:
   a. Fundamental Problems of Economic Planning
   b. Industrial location
   c. Land Policy
   d. Prices, Wages, Income and Productivity

Memberships in the above groups are made up of senior executives from business, education, government, and other institutions. Moreover, they relate to the problems and formulate policies to be implemented by business, education, government and other institutions with definite goals and milestones with time limits that provide for assessment of the effectiveness of their policies.

These barely visible emerging complexes are creating the need and perception for broader and more deliberately and skillfully designed goals and criteria for the formulation of institutions and their interrelationships. Current macro-engineering systems are no longer self-contained but are highly interrelated. This situation would seem to be tailor-made for a systems analysis; but it might be noted that in order to carry out a systems analysis, you first need a system to analyze. And as was previously indicated, neither policy centers nor society's problems have been systematized. While a partial solution includes the need for the newer institutional complexes that are yet to be clearly defined and structured, we must be aware that many of today's societal problems are recurring. This necessitates retaining many of today's institutions and their interrelationships. The newer institutional complexes are therefore, an added layer over today's existing institutions thus creating the need for an order of magnitude of management abilities that have not been required heretofore. What is more imposing is that such management has had no development to aid them in structuring the problems for solution other than those which can be extended from macro-systems. What this means for society is that:

1. Macro-systems present opportunities in the form of new products, new services and new markets for both the existing institutions and the emerging new institutions.

2. Government is both a generator and a consumer of the services and goods derived from the solutions of the skyrocketing demands (e.g., demands for minimum levels of housing, food, health care, education, highways, protection, etc.)

3. We can anticipate changes in public policy that provide means for attaining the desired partnership among government, industry, labor, and the public in strengthening the well-being of our nation.

Overview

The time is long past for the installation of common-sense methods for evaluating macro-projects and for the adoption of realistic guidelines to decision-making. According to William J. Jones, the evaluation process must consider the motivation for a large-scale project, its value to society as a whole, its practical risks, and the likelihood of changing or halting a macro-engineering effort once construction has been authorized. Government is often heavily involved, either through funding, regulation or guarantees, and the political process therefore plays a critical role. The evaluative process, if it is to be candid, must identify political forces supported by the special interests of various groups: the vocabulary of systems science should delineate--and not obscure--the underlying realities in a society where public decisions are responsive to constituency pressures.

William J. Jones

3. Macro-Engineering: How to Decide

The epilogue of Dividends from Space describes a letter from a nun to a U.S. space scientist asking how one could justify spending billions of dollars in space research while millions of children were starving on earth.

In his reply, the scientist compared the space budget with the far greater expenditures by Americans for cosmetics and tobacco and stated that space activities should be less vulnerable in the scale of values than lipstick and the kingsize filter-tip cigarette, and that even if the space program were to be dismantled, social problems such as hunger, crime, urban decay, pollution, etc. would not go away.

Macro-engineering projects are too often proposed as heroic enterprises and excluded from adequate evaluation by stating that conventional yardsticks used to evaluate and cost projects, and rules of management and budget control, do not apply. Because of the category into which the proposals fall, they will always be for the benefit of all mankind and the "benefits" are needed!

It is the intent of this paper to suggest that the time is long past for development of methods for evaluating macro-engineering projects and the early adoption of some guidelines in decision-making, even though they may be rudimentary and will require considerable improvement.

Macro (meaning large or extensive), as applied to engineering, can describe size, technical difficulty, time required for design and construction, initial costs in money, size of labor force, time to completion, magnitude of impact, etc.

Some projects are revealed as macro-engineering efforts only when viewed after completion. New York City, London, Rome, etc. fall into this class; they "grew" without a prior prepared detailed blueprint. Washington, D.C., Brasilia, and the proposed new capital of Nigeria, are examples of progressively greater degrees of city pre-planning.

"Hot" and "cold" wars involve macro-engineering. Such efforts are characterized as being based almost exclusively on the perceived comparative end-effectiveness of projects. Costs in dollars, environmental impact, etc., are relatively minor, if at all, considerations.

Appendices A,B,C, and D describe other macro-engineering projects. To fall under this classification, a project need not be large in itself. It can involve a very small effort and have a very large impact. "The Pill", incandescent electric light lamp, internal combustion engine, "Coca-Cola" are examples.

Macro-engineering includes all projects which have sizeable impacts on society, the economy, the environment, governments and lifestyles.

The classification of impacts into first order, second order, and higher order groups changes with time, political winds, and the interaction with other such projects and frequently singularly insignificant events, discoveries or ignorances.

A Definition of Macro-Engineering

I suggest that a project can be defined as "macro-engineering" if it falls under one of the following categories:

1) It entails government funding, or other government involvement (guarantees, special tax incentives, etc.) because of the magnitude of capital investment requirements, the extent of environmental impact, the time span to completion.

2) Uses exhaustible natural resources that may be

renewable but only after a long gestation period.

3) Large numbers of the population or particular segments of society will be affected.

4) Requires substantial participation of state governments, foreign governments.

5) Obligates the government to monitor, control, safeguard the products, plant or residue for long periods of time or in the event of failure of the private sector to do so.

Why Should We Be Concerned?

It is this author's suspicion that we have reached, or have even overstepped, the limits of our economic, manpower-management, and social tolerance limits.

There is pre-occupation with government funding of politico-economic innovation for achieving national goals: war, energy supply, transportation, space exploration, social welfare (health, recreation, etc.).

There is increasing national and international pressure for spectacular technical advances.

In the next 15 years we must mobilize as many raw materials as have been extracted during all of man's previous history on this planet.

Within the next 10-15 years we must design, manufacture, install and bring into full operation as much power production equipment as has been accumulated up to this point in our history.

A characteristic of technological advancement is a decreasing requirement for labor in production, more sophisticated scientifically-intricate production means and increasing attention of scientists to research and development. Every LDC wants to make a quantum leap into modern industrialized-nation status. The rising expectations of the masses of peoples of the Third World are encouraged and accelerated by television, movies, and newsprint. [The standards of living in established countries of the First and Second Worlds are threatened by shortages in natural resources and controls of supply in countries of the Third World.]

The commitment of capital to construct and tax support to operate, regulate, monitor and dismantle, strains the national economy, the social fabric, political security and the physical and biological structure of our planet and its atmosphere.

There is a centralization of decision making by persons whose accountability in time is much shorter than the time required for demonstration of failure or success; injudiciousness of the original decision becomes apparent only when the "ins" are "outs."

Motivation

The motivations for initiating macro-engineering projects may be classed under personal desire for power, conceit, religion, monetary profit, political stability or advantage, national pride, competition, growth, health, safety, etc., including "the good of mankind."

For a number of nations there is no demarcation between theological and political concerns--they are one. With others, political philosophy and action have replaced conventional concepts of religion without the population's being fully aware of the event. In many cases, engineering projects serve only to advance the established "religion" or provide more support for the prevailing "theology/politics".

The most powerful motives are of a competitive sort: "We want to come out on top." We want to be able to credit ourselves with first place, to be second is bad, and to be second-rate is intolerable. Motives such as the promotion of scientific, technological and economic progress are less compelling in political circles, though in other circles one or another of them may be dominant.

The competitive spirit and the desire to excel are important and an integral part of American life. There has always been an element of emotional commitment to big projects. If a project is seen as a challenge, the view is that it is good to excel, that it is good to test one's mettle against significant challenges.

Political Stabilization, Gains and Diversion

Large projects are apt to evoke domestic enthusiasm and national pride and awe and respect from the leaders of many of the nations of the world. They can be counted upon to gain international prestige and augment national pride with all shades of opinion, reflecting differing philosophies and sets of values across a broad spectrum. At one extreme, the development of a project is a technical problem resolvable through engineering and economic efficiency. At the other extreme, symbols for international cooperation and greater sociopolitical understanding and progress appear to dominate the evolution of a "high technology" program.

That leaders and political parties seek to promote their own political fortunes need not surprise us. The "ins" champion policies (goals and rules and methods for achieving them) with an eye to the next election or crisis. The "outs" look for weaknesses, failings and omissions. They try to devise alternative policies with public appeal.

Political concern does not focus primarily on the scientific and technological measures to implement a program often asserted in generalities.

We should be more curious about the values and interests -- the motives or reasons -- that inspire political behavior, especially in the field of international relations. To what extent is a project motivated by a desire to enhance American prestige over the world or to have one's name recorded in the annals of history?

The technical characteristics of the military arms industry have changed since 1955:

Salaried employees	25% to 46%
Engineers and Scientists	10% to 16%
Floorspace devoted to manufacturing	52% to 40%
Laboratories and Offices	18% to 28%
Manager, schedule controller, procurement and overhead services people	14% to 29%

The changes indicated above imply a parallel shift in the composition of many legislative constituencies. The political implications are obvious.

We must try to evaluate giant projects associated with modern warfare in terms other than profit and employment opportunities.

By enlisting private organizations in the performance of public functions, government involves them in politics and blurs the line between "public" and "private" industry. The private organizations, however, retain the advantages of private enterprise while serving the vital needs of the nation and still influencing their own futures. Hcw do we deal with this "grey area?"

Is it physically, politically and economically possible to build, operate, and maintain future macro-engineering projects in a common-sense manner?

The dilemma of the situation in which we find ourselves was expressed by a person most involved in the space program:

> "We need to have and understand nationally-accepted goals or purposes.
>
> How can we decide how important it is to spend, on an urgent basis, the very large sums of money required to put a man into orbit, etc., unless we have a pretty firm grasp of what the purpose behind the whole space effort really is?
>
> And yet, who knows the answers to this and many similar questions today? Who is thinking about them and doing something about developing some answers?"[1]

When a project is advocated, how do we determine:

a) that it will accomplish what its promoters claim;
b) what hidden advantages are there, i.e., advantages other than those suggested by the proposers;
c) what common yardstick of "values" can be used to compare one project with others;

[1] T. Keith Glennan, First NASA Administrator.

d) what are the "true values" of each of the several classes of costs.

Value

What is the value of a project? How do the values of one project compare with another's? How can one compare different values? How can one apply a number to a value? What is a value "worth" in dollars?

There is no absolute way of proving or disproving one or another goal value. Some have only temporary or future values. One project has values only when compared with the values of another project. Some values exist only at the sacrifice of others. To propose projects for security, prestige and pride is patriotic; to challenge projects that are purported to be for those ends is unpatriotic.

1) Technology is becoming more voluminous and more complicated.
2) Large-scale engineering projects have completion dates too far into the future to permit adequate assessment or no timetable at all.
3) Much new technology is so complex and the time span to stages which permit reasonable evaluation is so wide or indefinite that it is extremely difficult to anticipate how it will do its primary job and what its second-order consequences will be.
4) As our understanding of biological, ecological, economic and social processes improves, as we observe and realize the immediate consequences, we have an obligation, under our planet stewardship responsibilities, to evaluate, to the best of our abilities, the second-order consequences of all our actions, and to include their costs, monetary and otherwise, in our analysis and decision process.

Exploration of space has been compared with the voyages of Columbus who was funded by Queen Isabella who, in turn, was funded by peasants.

There were many other explorations which we do not hear about. Expeditions that were very expensive to initiate and ended in disaster.

Proponents of a project will always compare it with a highly successful one. The evaluation should compare it equally with those that were unsuccessful.

Many of the programs being advocated at this moment (for example, the world-wide conversion to coal as an energy source) have planetary phenomena intrinsically associated with them. Since the impacts will be global, control and management are likely to be effective only if they can be organized on a world-wide scale. Full achievement now or within a foreseeable future is not possible; nevertheless, even partial consensus is desirable.

Most of the major public engineering expenditure decisions have been characterized by "muddle through" and "rule of thumb". Public money has been lavished often on "popular" projects with a very hazy idea of the return to be expected and even the extent of all construction costs and obligations once completed. Those who object are stigmatized as "wild-eyed idealists" or "vested interests."

There is continuing need to discover and explain the costs and benefits and to see that our social control system helps resources flow into the appropriate channels.

Dorfman warns, cost-benefit analysis may help but the art can be likened to the problem of appraising the quality of a horse-and-rabbit stew, the rabbit being those consequences that could be measured and evaluated numerically, and the horse "the amalgam of external effects, social, emotional and psychological impacts and historical and aesthetic considerations that be adjudged only roughly and subjectively." The horse is bound to dominate the flavor of the stew; meticulous evaluation of the rabbit is hardly worthwhile.

There is a major problem in evaluating macro-engineering projects (MEP). That is, the quantification of "externalities" and "group wants".

Non-compensated external costs must be addressed. "Group wants" either must be accepted as is or made to be founded on accurate and complete technical information.

There are inherent hazards in leaving of

decisions about MEP's in the corridors of political power. The establishment of economic and social priorities by Congress reflects primarily political pressures by vested interests. If one does have faith in the conventional political process, can we create tools which the Congress and government officials might be obligated to employ to maximize objective evaluation and action?

Is there a social rate of discount to be applied to MEP's because the magnitude of the investment is a sizeable portion of the GNP and the completion date of the project falls outside of the period of responsibility and accountability of the proponents? Can the MEP be analyzed within or in isolation from the rest of the economy? Can a decision to continue or discontinue an MEP be made in light of anticipated conditions at any point after it has begun?

The summary question is -- how does one determine the cost and benefit streams of government-sponsored macro-engineering projects?

The determination of rates of return on capital acts as one of the prime movers in resource allocation in the private sector of the economy. The government must develop a procedure for testing the "soundness" of macro-engineering projects, "soundness" being the desires of the public after being informed of all aspects.

Risk

A fundamental element of social control is the necessity for full and candid disclosure and discussion of potential costs and benefits. One cannot expect that those who are sponsoring the development of an MEP will indulge in listing its undesirable social consequences since an inherent feature of the promotion process is to minimize these consequences and to argue that they can be technologically overcome.

Investment by the private sector considers risk, the project will not "pay-back" as expected.

To what extent do the government-contract funded projects reflect consideration of risk? Are the proposers subject to penalty if the pay-back is not as promised?

With an MEP, where performance can only be measured many years after the start of the project and tens of millions, or in some cases, billions, of dollars have been invested, there is no "inordinate" risk perceived by the proponents.

Conclusion

Improving the process by which any sizeable federal program is conceived, evaluated, approved, developed, financed, managed and terminated is a challenging task.

The complex nature of our political system, the stubborn resistance to change inherent in a giant governmental bureaucracy, pose formidable obstacles.

With MEP's, the attendant cost of hundreds of millions and billions of dollars, and the time from conception to realization lead one to wonder whether anything at all can be done.

Nevertheless, the magnitude of the irreversible commitment to macro-projects which will steer policies for a few generations, and our involvement in the internal detailed affairs of other countries and the relationships between them and us, make it essential that we allocate a considerable portion of our attention and efforts to the task.

We must be concerned not only with efficiency (adherence to budgets and completion dates) but also to value (value related to all other economic and social needs and desires) and the objectives (sub-goals) as related to reaching long-term mission or direction for public policy.

The goals of public policy must be developed, specified and ratified by the political process as an expression of the people's will.

In order for that will to result in effective action, government officials and the public at large must participate in a continuing process of self-education.

Objectives and even goals must be revised as

more is learned about their costs and inter-relationships.

There must be an awareness of and comparison with alternatives to a proposed project. Valid analysis requires fundamental research and experimentation on relationships between means and ends, results and costs.

MEP's have output or cost implications that extend significantly beyond the federal government's one-year operating and budgeting period, more often than not, beyond the tenure of political personnel and frequently into future generations.

In the execution of the program, responsibility must be assigned at all stages for costs, and adherence to completion schedule.

Effective program progress evaluation requires standards of measurement, proper relative weighting of the several criteria and data.

A comprehensive technology assessment must precede the proposal of a project. Technology assessment must be conducted at frequent "milestones" during construction and throughout its use so that modification or termination is instituted at the proper time.

a) Imbalances that arise during and after MEP's are correctible only with great difficulty _ex post facto_ actions. Because of the magnitude of any or all elements, risks are prone to become irreversible.
b) If it is determined or suspected that an MEP endangers economic growth, environmental stability, social quality of life, or turns out to constitute a major menace of depletion of natural resources, will inertia prevent modification of unpleasantnesses in time to avoid severe exacerbation?
c) Socially responsible management of an MEP is virtually impossible. Too many facets of society are affected, some positively, some negatively, in major ways. More important, the Heisenberg uncertainty principle -- the precept that the accurate measurement of an observable quantity necessarily produces uncertainties in one's knowledge of the values of other observables -- applies in social behavior.

Government must learn to distinguish between:

1) regulations which protect the public's financial and other interests, and;
2) regulations which result in a loss of industrial incentive, creative ability and responsibility.

It must improve its relationship with industry by increased emphasis on competitive award of contracts and cost-plus incentive-fee contracts.

It must increase pressure for industry to assume a greater share of risk.

Government must pay increased attention to relatively unsophisticated components and systems which are in small sizes, impacts and dollar value. An outline of a partial test or evaluation of an MEP follows:

1) Need or desire: who
   a) wants
   b) advocates
   c) profits
   d) loses

2) Has the general public been given a chance to hear open debate, and an opportunity to register approval or disapproval?
   a) Alternatives. What can be done to accomplish same <u>end</u> result?

3) How much of "need" or "desire" is based on:
   a) transient conditions
   b) "soothsayers" prediction vision of the future
   c) creations by vested interests.

4) Who finances? and what are the risks during
   a) planning
   b) construction
   c) operation
   d) dismantling
      -- because of failure
      -- outliving usefulness

5) Is cancellation of project possible and where along its life do the possibilities exist?

There is a technological race among the powerful or would-be powerful, and an increasing gap between the "haves" and "have-nots."

There is a race between what might be called the technological factors of equalization and the countless disparities and inequalities -- military, economic, political, ethnic, ideological -- which distort this revaluation and often put it in the service of the most powerful and of the most dissatisfied.

Not all MEP's nor all aspects of a particular project can be evaluated in dispassionate, objective terms. There are casual conditions that are beyond human control, but the conscious reasons that people entertain concerning the wisdom of this or that action should be determined and understood.

Evaluation requires the constant presence of awareness that the programs are always, some to a greater degree than others, self-serving. In our society, we encourage competition and achievement by offering personal incentives (money, fame, power, etc.).

The programs involve tens of years so that the proponents' identities are lost. There is absolutely no liability for failure, no one and no organization is accountable. The original promises are forgotten.

The scientist stands to get research funds, the university anticipates grants, non-profit research organizations want contracts out of which they can pay high salaries, the trade union wants to keep up employment, the business concern wants profits, the trade-journal caters to the complex of readership, the congressman seeks re-election, the promoter (governor, elected official etc.) wants to bring more wealth to a state or region, and the U.S. President wants immortal fame -- beginning within his tenure in office.

Overall there is a very definite limit to growth, and within that overall limit, a limit to rate of growth. World demand is not only for energy, but for food, forest products, minerals, fresh water, marine protein, skilled labor and so forth, as a function of rising expectations, rising affluence and rising population numbers. The technologies that underlie our economic system evolved in a situation of relative resource abundance. The real dilemma that we face is the need to impose a rational and conscious choice upon the limited alternatives which confront us.

"For which of you, intending to build a tower, sitteth not down first, and counteth the cost, whether there be sufficient to finish it?"
(Luke 14: 28)

Appendix A
Trans-Planetary Subway Systems

Robert Salter of the Rand Corporation suggested, at last year's annual meeting of the AAAS, the adoption of the "Planetran" concept, a subway system moving at thousands of miles per hour, capable of crossing the United States in an hour or so. It was proposed as a possible alternative to the anticipated over-filled and hazardous airways/airports of the year 2078.

Who but the government could possibly begin to even explore the feasibility of such a project? Once commenced, who would be powerful and secure enough to stop funding if it seemed not to be a viable solution to the problems? Once the feasibility study of this concept is begun, what chance would there be for any other project to replace it even if the alternative were more promising? How close to the actual costs can estimates be? What could the government do if it found that costs were exceeding estimates?

How do we compare it with other proposed solutions? Not only technical characteristics but human preferences, prejudices and expected life styles which could change several times over the period required for project completion will have to be taken into account.

Appendix B
Erie Canal*

The Erie Canal was the work of that remarkable generation in America which made the period between 1815 and 1860 an age of great national expansion. It was a bold scheme designed to bring the Mississippi through a northern waterway, on American soil, to the growing metropolis at the mouth of the Hudson River. The supporters of the idea stated that the result would be national growth, strength and prosperity.

Travel between the west and the northeast coast cities was by poor roads, or in part, along the St. Lawrence River, which was closed by ice for a number of months and involved contact with the Canadian government and terminated in Canadian cities.

It cost $2.00 to send a barrel of flour 130 miles overland and the same barrel could go by water from Albany to New York City for 25 cents, a distance of 160 miles. A waterway between Buffalo and Albany would provide cheap, safe, "American-owned" transportation between the West and the East. A cannon worth $400 in Washington cost $2000 to transport to Lake Erie. At the Niagara frontier there was a foreign power, Canada, which controlled the only outlet of the Great Lakes to the ocean.

The Commission charged with evaluating the "worth" reported favorably and included the statement.... "After a lapse of two thousand years and the ravages of repeated revolutions...this national work shall remain...."

The report theme was: the potential productivity of the uninhabited Western lands, the pressing need for communication, the visions of private and public gain, the fear of Canadian rivalry. The project was funded by New York State and administered by a board of Commissioners, all with political ambitions.

---

* See the extended discussion of this project by Jeanne Krause in Chapter 7 of this volume.

It was estimated that the canal could be completed in ten or fifteen years at a cost of six million dollars. The project, with its thousands of jobs, was tailor-made to augment the power of the patronage in New York. It cost a little over seven million dollars, or about 16% more than estimated.

The canal did provide the promised communication with the West. Western products were exported through the canal to New York City in amounts greatly in excess of the original estimates. Merchandise reached the mid-west and the northwest from the East Coast through the canal. Emigrants traveled the route by the thousands and contributed to the rapid development of the West.

With the proliferation of railroads, the worth of the canal began to decline. It is non-existent today. But it exceeded its predicted contribution to the development of the West. The economist, W.W. Rostow, cited it as the principal contributor to the "take-off" stage of national economic growth in the 1840's.

Appendix C
Military Macro-Engineering

U.S. Military MEP efforts during both war and peace have been described as cooperative arrangements between the military, the arms industry and--frequently--political parties. It has been argued that military projects must be viewed with great suspicion and concern.

The proponents are described as having well-established vested interests and join in opposition to any form of disarmament plan, promote new and more expensive projects and sometimes appear to be engaged in an orchestrated "see-saw" escalation of an arms race with perceived enemy countries.

If there is any basis in fact for these charges, one wonders if it is at all possible to evaluate projects in an atmosphere devoid of profit motive, opportunities for the employment of scientists and engineers, much less the eternal issues of personal pride, ambition and the egos of political leaders.

One is tempted to ask whether France's spectacular commitment to the Maginot Line could not have been usefully replaced by the building of the armoured formations sought by iconoclasts such as (Colonel) De Gaulle!

Appendix D
"Lessons Learned from Constructing the Trans-Alaska Oil Pipeline"

In a 1968 feasibility cost study a private group estimated that an oil pipeline system from Prudhoe Bay to Valdez, Alaska, would cost $1.046 billion for a 1.2 million-barrel-a-day capacity. On the basis of this estimate decision makers calculated the cost of a barrel of oil delivered to the "lower 48", and concluded that the benefits (including environmental impacts) were positive and that it would be in the best interests of the public to grant rights-of-way, through federal and state-owned land, to a private company (Alyeska Pipeline Service Company) to design and construct the pipeline system.

Shortly after pipeline construction began in 1975, the company established a base control budget of about $6.4 billion.

By December, 1977, at the completion of the line, the cost was $7.9 billion.

It is argued that this project was privately financed. The federal and state governments, on the basis of cost-benefit analysis using the original estimate, granted rights-of-way from the Arctic to the Pacific Ocean.

The General Accounting Office, at the request of the Senate Committee on Energy and Natural Resources, reviewed the project and recommended that the following should apply to similar future projects:

a) First and subsequent cost estimates should be viewed with skepticism.
b) As much site-specific data as is economically practicable should be obtained.
c) Technical and geological uncertainties should be thoroughly investigated.
d) Government approval should be contingent on detailed planning for management control, including budgetary controls.
e) The Alaska Natural gas pipeline project's expenditures should have an ongoing government audit to protect the public interest.

Appendix E
Motivations for Adoption of a Project

Motivations for adoption of a project can be classed as follows:

a) Military Security:

1) immediate: communication, navigation, transportation, and/or other basic capabilities
2) potential: new technology, weapons, etc.

b) Peace

Mitigation of military irritants or increased supply of shared geography, resources or opportunities

c) Progress in Science and Technology

1) The desire for knowledge
2) Technological progress (technical efficiency)

d) Economic and Social Progress

1) Direct: new products for home, industry (created desires, created markets)
2) Spill-over: new industries
3) Social Effects: health and safety

Miscellaneous Possibilities (Socio-Economic Mobility and Fantasies)

e) National Prestige

What, however, is the meaning and importance of "prestige"?
In whose minds and to what ends?
For what?

f) National Pride

1) achievement
2) political (party) security
3) challenge from others (imagined or real)

g) Special Interest and Ulterior Motives
(Private and expressed)

1) Personal incentives
2) Private advantage
3) Catering to constituencies
4) Keeping going concerns going
5) Serving partisan interests

Bibliography

"Dividends from Space"
F.T. Ordway, C.C. Adams, M.R. Sharpe
Thomas Crowell Co., N.Y. 1971
A "documentation" of the benefits of the Space Program

The Industry-Government Aerospace Relationship
SRI Project No. IS-4216, 1963.
Stanford Research Institute, Menlo Park, CA

The Politics and Technology of Satellite Communications
Jonathan F. Galloway
D.C. Heath & Co., Lexington, MA 1972

"A Technology Assessment Primer"
L. Kirchmayer, H. Linstone, W. Morsch, editors
No. 75 HJ 3137-7TFA Inst. of Elec. and Electronic Engineers, N.Y., NY. 10017

"Measuring Benefits of Government Investments"
R. Dorfman (ed.)
Brookings Institution, Washington, DC 1965

"Lessons Learned from Constructing the Trans-Alaska Oil Pipeline"
Report of the Controller General to the Congress
EMD-78-52,
U.S. GAO, Washington , DC 6/15/78

Overview

Macro-engineering programs and projects pose difficult technical and managerial challenges at every stage of their development. C. Lawrence Meador and Arthur C. Parthé, Jr., argue that these challenges derive from several characteristics of macro-engineering projects and programs that are not found (or are only found in relatively minor ways) in conventional projects and programs. Macro-engineering projects often require significant quantities of public and private sector resources that must be committed (and thus put at risk) over relatively long time frames. Difficult technical and scientific problems must frequently be solved before the projects can be accomplished. This may result from the raw size of structures to be put in place, from systems complexity or from the emphasis on developing previously unknown technologies. Finally, macro-engineering programs may, and in fact often do, have profound socio-cultural impacts in the societies which develop them. This raises a host of issues including environmental implications, legal and regulatory factors, economic and political effects, to name only a few. Unfortunately, existing managerial systems, methodologies and frameworks do not address this portfolio of multi-dimensional objectives and constraints in an effective manner. A framework is presented which characterizes macro-engineering programs and projects from a managerial policy, planning and control perspective. From this framework a multi-dimensional and relativistic project state space is developed in which to portray analytic representations of macro-projects.

C. Lawrence Meador and Arthur C. Parthé, Jr.

4. Managing Macro-Development: Policy, Planning and Control System Implications

Introduction

Macro-engineering programs and projects pose difficult technical and managerial challenges at every stage of development including planning, research, analysis, design, implementation, testing, evaluation, demonstration, and operational deployment. These challenges derive from several characteristics of macro-engineering projects and programs that are not found (or are only found in relatively minor ways) in conventional projects and programs. Macro-engineering efforts require that huge quantities of public and private sector resources be committed (and thus put at risk) over relatively long time frames.

Macro-engineering projects often require the solution of difficult technical and scientific problems before they can be accomplished. This may result from the raw size of structures to be put in place, from systems complexity of interacting components and subcomponents, or from the emphasis on developing previously unknown technologies.

Finally, macro-engineering programs may, and in fact often do, have profound socio-cultural impacts in the societies which develop them. This raises a host of issues including environmental implications, legal and regulatory factors, economic and political effects, to name only a few.

Unfortunately, existing managerial systems, methodologies, and frameworks do not address the portfolio of multidimensional objectives and constraints which we have alluded to in an effective manner. At this stage, we need a framework and a methodology to:

- Facilitate better policy analysis, planning, and control of macro-engineering projects and programs,
- Address non-financial and non-technical issues (e.g. legal, political, environmental, etc.), and
- Create an audit trail for the examination and resolution of problems and decisions arising during development of projects.

Further, we need to provide project and program managers at all levels with better tools and techniques for effective technical and managerial planning and control.

Decision Support Systems

Over the past several years there has been a dramatic trend toward providing managers and technologists in complex problem and decision environments with computer-based systems which help to organize, store, process and present information relevant to planning and control issues which they face. Recently, a new trend has emerged which focuses on difficult semi-structured problem environments where programmable solutions are not available which "solve" the problems addressed. In some of these cases a computer aid can be used in various modes to support the problem-solving and decision-making activity that is required. Scott Morton calls this new trend in the development of computer-based decision aids the Decision Support System approach.[1] We believe that decision support systems have exciting potential for addressing many of the difficult macro-engineering planning and control issues which we allude to above.

The balance of this discussion presents the rudiments of a framework for managerial planning

---

[1]See, for example, Keen, P.G.W. and M.S. Scott Morton, Decision Support Systems, Addison-Wesley, Reading, MA, 1978.

and control of technology development in macro-engineering projects. The major dimensions of the framework include issues related to project cultural impact potential, technology development difficulty, and public/private resource requirements. This framework is implemented in a new emerging project planning and control decision support system called MACRO-PLANNER which is currently under study and development at MIT and the Charles Stark Draper Laboratory, Inc.

The Framework

Figure 1 presents a three-dimensional visualization of what we shall call macro-engineering state space. It can be thought of as an abstraction of the key dimensions through which macro-engineering projects or programs must be evaluated. We can further refine these dimensions by listing their key components in the following ways:

Cultural Impact Potential
- Environmental Implications
- Political Effects
- Life Style Impacts
- Legal and Regulatory Results
- Changes in Attitudes, Perceptions, Options and Opinions

Technology Development Difficulty
- Fundamental Science Barriers
- Systems Complexity-Size/Structure
- Interdependency Problems
- Modelability - Physical and Analytical
- Empirical versus Theoretical Emphasis

Public/Private Resource Requirements
- Manpower - Technical, Managerial and Other
- Funding from Public and Private Communities
- Materials, Supplies, Critical Raw and Manufactured Goods
- Governmental Inspection and Regulation Requirements
- Energy Demands

A macro-engineering project can be thought of as "extended" in macroengineering state space as a function of the magnitude or the degree to which individual components of the state space are impacted. Figure 2 suggests how three macro-

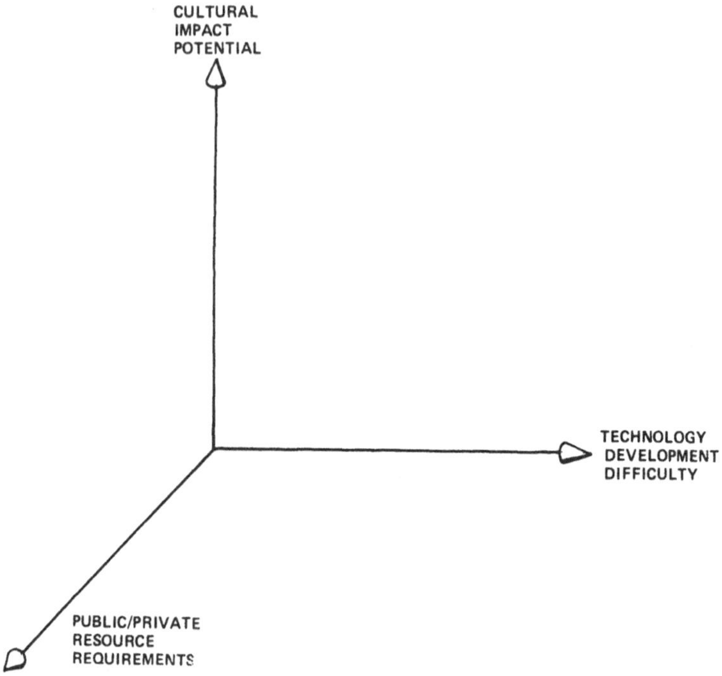

Figure 1. Graphical representation of macro-engineering state space.

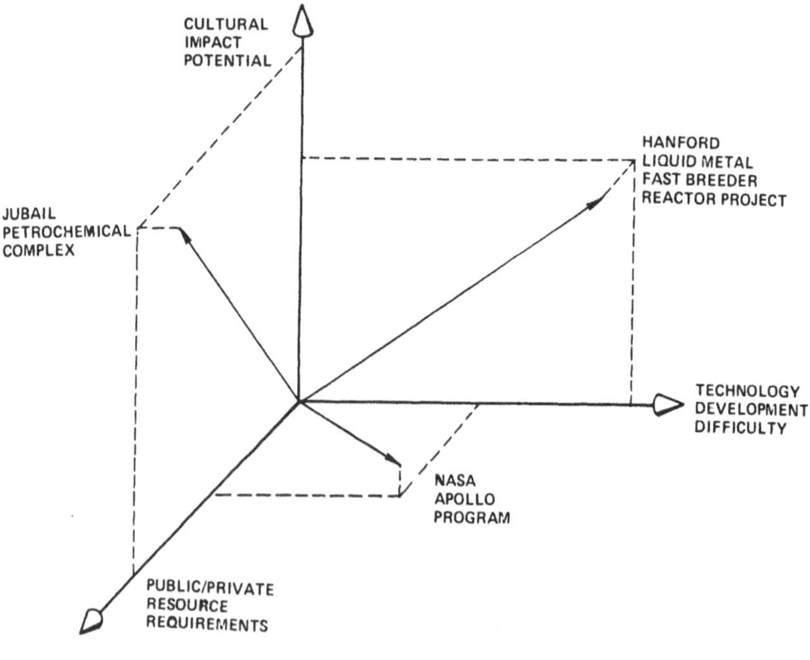

Figure 2. Comparison of macro-engineering efforts.

engineering projects might be viewed along these dimensions in comparison to each other. The dimensions of cultural impact potential, technology development difficulty, and public/private resource requirements are only relevant when considered in the relativistic context of the socio-cultural milieu in which they are evaluated. The building of the Pyramid to the Sun in the Valley of Mexico by the Teotihuacan or the construction of the astronomical observatory in Chichen Itza on the Yucatan peninsula by the Maya must certainly be viewed as macro-engineering efforts at their time and place of deployment. The Teotihuacan pyramid was equivalent in relative socio-cultural impact and resource requirements to the present day Saudi Jubail petrochemicals complex. The Mayan astronomical observatory required resources and technological or scientific accomplishments equivalent in relativistic terms perhaps to the U.S. Apollo program.

The Information Hierarchy

To be useful for developing meaningful and relevant insights and guidance toward planning and controlling the deployment of macro-engineering programs and projects, a framework must be sensitive to the type, quantity, and characteristics of information appropriate to different parts of a macro-engineering organization. Figure 3 illustrates the information hierarchy pyramid which characterizes the policy and strategic planning, program management control, and project operational control functions necessary in a macro-engineering organization. The implication of the pyramid is that a great deal of data (and people and organization units) exists at the lowest operational levels but that this information is only necessary and appropriate in highly aggregate form for decisions of a strategic or policy type at the top.

Figure 4 demonstrates the very important principle that different managers within a macro-engineering organization may have dramatically different information needs depending on their level and type of responsibility, organizational location, and personal managerial "style." In attempting to develop a macro-engineering planning and control decision support system, it is essential that such differences be taken into account. The implementa-

Figure 3. Information hierarchy.

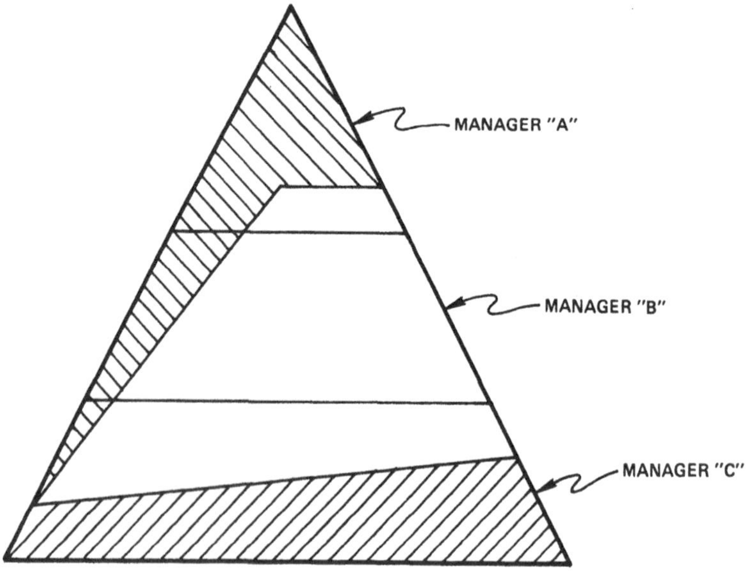

Figure 4. Differing managerial information requirements.

tion of multiple models and information processing options for addressing problems from different perspectives is a useful and relevant way of accomplishing this objective in a decision support system environment. More will be said on this issue as we describe our own efforts in developing the MACRO-PLANNER system described later.

Structured Versus Unstructured Decision Problems

An important though somewhat hazy distinction can be made between so-called programmed and nonprogrammed decisions:[2]

> Decisions are programmed to the extent that they are repetitive and routine, to the extent that a definite procedure has been worked out for handling them so that they don't have to be treated "de novo" each time they occur... Decisions are nonprogrammed to the extent that they are novel, unstructured, and consequential. There is no cut-and-tried method of handling the problem because it hasn't arisen before, or because its precise nature and structure are elusive and complex, or because it is so important that it deserves a custom-tailored treatment. By nonprogrammed I mean a response where the system has no specific procedure to deal with situations like the one at hand, but must fall back on whatever general capacity it has for intelligent, adaptive, problem oriented action.[3]

Problem environments can be characterized as structured or unstructured by the extent to which programmed or nonprogrammed decision procedures apply to them. Most interesting management decision problems appear to fall into the relatively

---

[2]Meador, C.L. and D.N. Ness, "Decision Support Systems: An Application to Corporate Planning," Sloan Management Review, Winter 1974.

[3]Simon, H.A., "The New Science of Management Decision," New York: Harper & Row, 1960.

CHARACTERISTICS OF INFORMATION	PROJECT OPERATIONAL CONTROL	PROGRAM MANAGEMENT CONTROL	PROGRAM/PROJECT POLICY AND STRATEGIC PLANNING
Source	Largely Internal	⟵⟶	External
Scope	Narrow	⟵⟶	Very Wide
Level of Aggregation	Detailed	⟵⟶	Condensed
Time Horizon	Historical	⟵⟶	Future
Currency	Highly Current	⟵⟶	Quite Old
Required Accuracy	High	⟵⟶	Low
Frequency of Use	Very Frequent	⟵⟶	Infrequent

Figure 5. Information requirements by decision category.

unstructured category. A typical attitude which has prevailed in the past is that substantially unstructured problems are either too trivial to require decision support or so complex that it is impossible to make decision support system technology relevant. A more useful design hypothesis is that such problems are, in general, neither trivial nor impossible; but they are difficult to solve.

It may be constructive to attack problems in terms of subproblems and subdecisions that can be supported rather than solved by a computer-aided decision process. Aspects of the total problem are sought where structure is recognizable, and the computer is used to improve and mechanize those aspects. Such subproblems may include comparison operations, production of graphs or other data presentation, and arithmetic or primitive logical operations. An understanding of common subproblems helps separate real problems into more manageable parts. The division of a problem into subproblems accomplishes two distinct things:

1) Problems rarely look alike, but they often share common parts which can effectively be operated on.

2) The user is helped to structure his thoughts in an improved and helpful way by suggesting what kinds of subproblems might be particularly useful to attack.

These principles are important underpinnings to the conceptual development and realization of decision support systems.

Information Characteristics

Figure 5 delineates another important set of design constraints for consideration in the development of a macro-engineering project planning and control decision support system. It is important to recognize that information characteristics vary dramatically by type and scope according to which decision environment is being addressed.[4]

[4]Gorry, G.A. and M.S. Scott Morton, "A Framework for Management Information Systems," Sloan Management Review, Fall 1971.

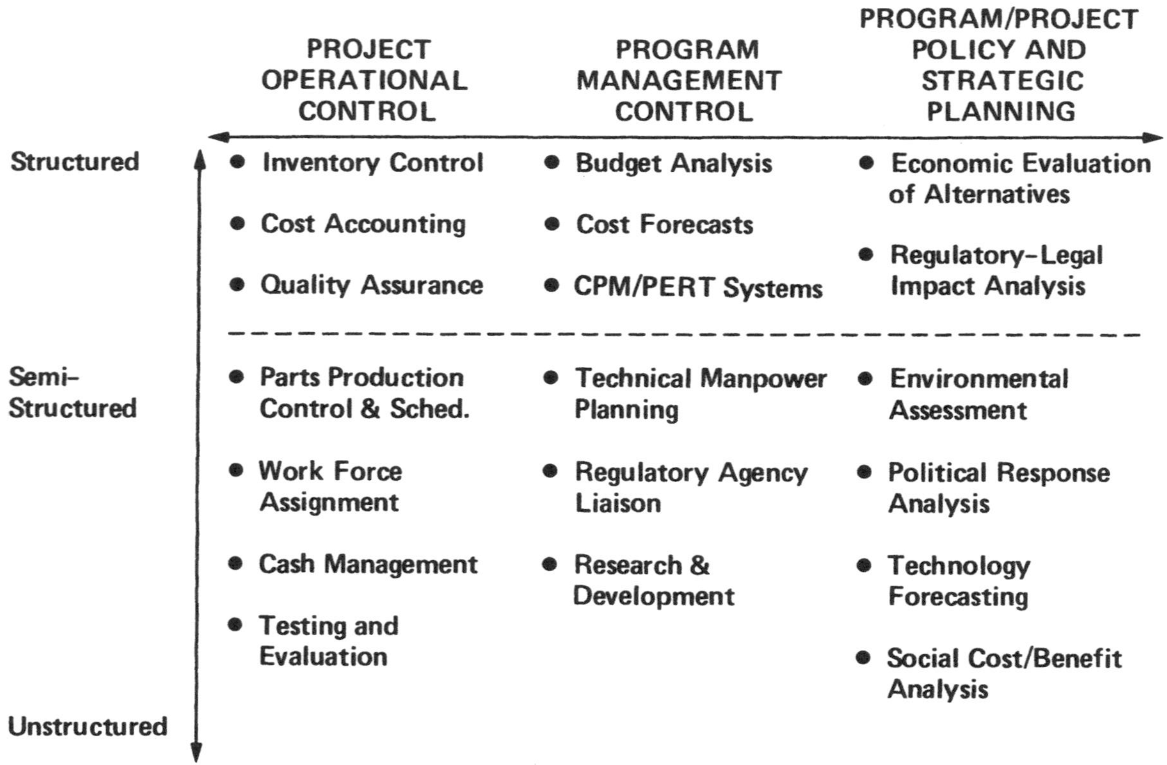

Figure 6. Macro-engineering management decision framework.

Earlier we stated that level of aggregation of information was quite detailed at the project operational control level but very condensed at the policy and strategic planning level. Required accuracy may be extremely high for operational control problems whereas a "reasonable guess" or a good estimate may be quite adequate at the strategic level.

From Transaction Processing to Decision Support

In the early development of the technologies of computer-based information systems, the field was led by an inward direction of concern with the hardware and software technology. This direction has resulted in a long series of orders of magnitude improvements in the processing efficiencies of new systems during the 1960's and 1970's. It can reasonably be expected that this progress should continue into the future.

There has also been a great deal of emphasis on the automation of relatively primitive, highly structured, previously manual tasks in areas such as personnel records, order entry, inventory control, and other accounting applications. Improvements in computer hardware and software technology have dramatically increased the <u>efficiency</u> of such transaction processing systems in recent years.

In Figure 6 we present a perspective on the macro-engineering management decision environment which we have characterized on the continua of structured to semistructured to unstructured decision problem type; and project operational control to program management control to program/project policy and strategic planning decision categories.[5] We have mapped into these two dimensions several specific tasks, applications, problems, and functions dealt with by macro-engineering technologists and managers. Inventory control, for

---

[5]This framework is a modified extension of early work on a general managerial information framework by Gorry and Scott Morton. See Gorry and Scott Morton, op cit., p.62.

The decision support system approach implies:

- Consequential problems in fuzzy, semistructured environments
- No programmed procedure applicable
- Requires managerial judgment, experience, intuition
- Tradeoff efficiency versus effectiveness
- Human interaction with computer
- Emphasis on supporting decision-making — not solving decision problems via the machine

Figure 7. Computer-based decision support system.

instance, can be viewed as a highly structured problem to be dealt with at the operational control level. At the other end of the spectrum we have tasks such as environmental assessment, technology forecasting, and social cost/benefit analysis which are relatively ill-structured and are dealt with at the policy and strategic planning level.

It is important to note in this framework that most of the computer-based information systems brought to bear on this problem context in the past have been relevant principally in the upper left quadrant of our figure. That is, the computer aids have been of the earlier transaction processing orientation which was concerned with highly structured operational control tasks such as inventory, cost accounting, quality control, reporting, etc. The decision support system approach, which we advocate here for macro-engineering project planning and control, has the effect of moving the context of concern from the upper left to the lower right quadrant of the figure. There is an emphasis on tackling some of the hard semistructured and unstructured policy and strategic planning issues. This approach requires us to start worrying about and attending to the problem of improving the effectiveness of organizational decision makers in macroengineering contexts who are users of computer-based decision aids. This implies that less time be spent worrying about improving the efficiency of computer hardware and software. The approach also leads us to emphasize a user-oriented process for designing the interface between decision maker and computer. Many macro-engineering technologists and managers may not be conversant with a computer programming language. Thus, we may, in many cases, be led to design a human-computer interface with natural language characteristics. This will facilitate the iterative problem solving and exploratory data analysis processes which seem to be important in rich, consequential, semistructured problem domains. Figure 7 summarizes many of the characteristics relevant to decision support systems in this context.

In the next sections, we develop a project modeling technique, the project management state space approach PMSS, and describe a project planning and control decision support system called

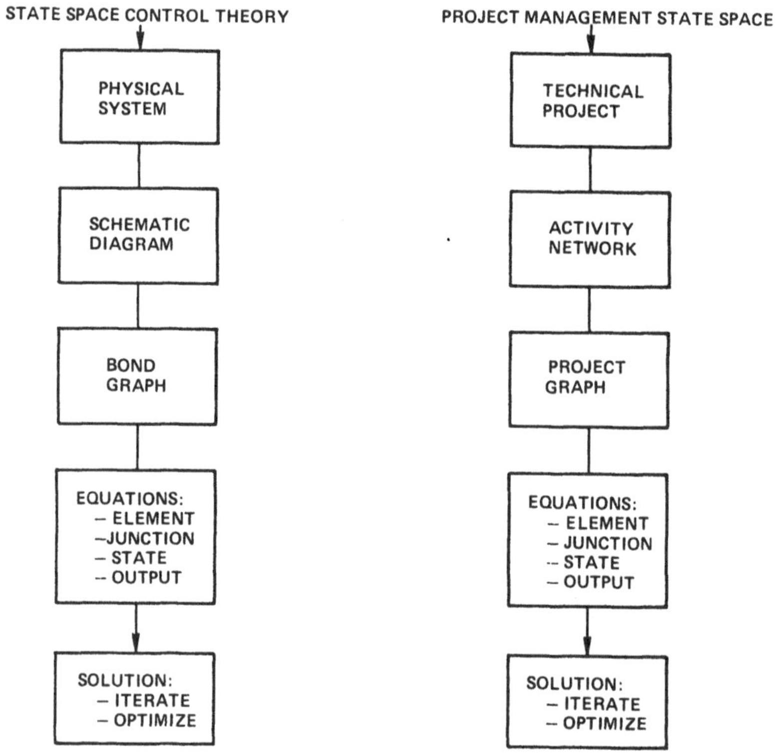

Figure 8. Parallel development and implementation of PMSS.

MACRO-PLANNER which we are utilizing as an experimental test bed for the modeling technique.

<u>The Need for a Model</u>

The managerial decision-making process for macro-scale project personnel can be enhanced through appropriate utilization of decision support aids. One such aid being developed by the authors is a state space project modeling system. The motivation for developing this tool is the need for having available an accurate and relevant analytic representation of the project environment. This permits experimentation with optimization of project structure, analysis of sensitivities to and consequences of various perturbations, as well as other categories of "what if" interrogations. The computational and structural capabilities of state space control system analysis techniques have been found to be adequate for synthesizing macro-scale projects. The appropriate analogies for linking control system concepts with the project environment to justify adoption of current state space concepts are developed in an earlier discussion of this issue.[6] This is illustrated conceptually in Figure 8.

The graphical representation of important project variables has received considerable attention over the years. PERT(Project Evaluation and Review Technique) charting is still considered a valuable representation system. Here the project is represented as a series of line segments that correspond to initiation and completion of that activity. These activity line segments are interconnected in a sequential fashion that reflects their chronological ordering. This technique has been of most value in a project planning context.

A similar graphical technique was defined in reference 6. It was noted that the bond graph method for diagramming engineering systems had

---

[6]Parthé, Arthur C., Jr., <u>Application of Modern System Analysis Theory to Large-Scale Engineering Development Projects</u>, Second Lawrence Symposium on Systems and Decision Sciences, Berkeley, CA (Oct. 1978).

certain similarities to project (PERT) networks. An activity graph project modeling technique has been defined and is derived from both the PERT project activity network approach and the engineering bond graph modeling technique. This facilitates a convenient and tractable coupling of the project environment with state space control systems theory.

Project State Space Concepts

The concept of multidimensional space has always had a particular intrigue and romance that alluded to hidden promises. A project analyzed in this framework is conceptualized as "originating" from the origin of the requisite coordinate system. As time elapses and resources are expended, the project moves along a trajectory through "project state space." The coordinates of this path are the minimal set of parameters to uniquely specify the project "location" at any given moment. These parameters include project cost, schedule, technical performance and other related quantities of interest. The difficulty in implementing these concepts was the lack of a specific scheme for tying this all together analytically. It is with the marriage of state space control system analysis and the conceptualized project state space that a unified and tractable methodology has evolved.

Project Management State Space/PMSS

Project Management State Space (PMSS) concepts embody a methodology and analysis framework that utilizes the resources of the digital computer. In a supportive role, PMSS can assist project management in planning and control decision making. It deals with the semi-structured project management details that require management judgement. PMSS is designed to enhance this process by making relevant information available on an as-required basis. The analytical framework for PMSS is derived from state space control system theory.

System Modeling

System representation in state variable form is given in the most general form by the following system of equations:

$$\underline{x}(t) = \underline{A}\underline{x}(t) + \underline{B}\underline{u}(t) \quad (1)$$

$$\underline{y}(t) = \underline{C}\underline{x}(t) + \underline{D}\underline{u}(t) \quad (2)$$

where

$\underline{x}(t)$ is an n-dimensional state vector,
$\underline{u}(t)$ is an r-dimensional control vector,
$\underline{y}(t)$ is an m-dimensional output vector,
$\underline{A}$ is an nxn system matrix,
$\underline{B}$ is an nxr control matrix,
$\underline{C}$ is an mxn output matrix.

The added term $\underline{D}\underline{u}(t)$ allows for a direct coupling of the output and input. Equation (1) is a set of n first order differential equations and in control is referred to as the plant equation. Equation (2) represents a set of m linear algebraic equations and is referred to as the output expression. The $\underline{x}(t)$ vector has as components the variables that uniquely define the state space location and trajectory of the system. Analysis of the equations allows for observation and optimal solution of the control system being studied.

Optimal System Performance

In considering optimal performance it is necessary to specify a method for determining the quality of performance for the system under consideration. The current approach is to specify an integral performance index of the following general form:

$$PI \int_{t_i}^{t_f} L(\underline{x},\underline{u},t)\,dt \quad (3)$$

The intent here is to achieve optimal performance over the interval from t_i to t_f by minimizing or maximizing, as appropriate, the performance index PI.

System Modeling with Bond Graphs

The accuracy and ease of analytically assessing the performance and control of physical systems is dependent upon the specific modeling techniques employed. Methods that are either too simplistic or at the other extreme, overly complicated, are generally not productive for the analyst.

Table 1a. Single port elements.

Systems	Variables		ONE PORTS		
			R	C	I
	effort	flow	resistor	capacitor	inertia
			$e = Rf$	$e = \frac{q}{C}$	$f = \frac{p}{I}$
	e	f	$\xrightarrow[f]{e} R$	$\xrightarrow[f = \dot{q}]{e} C$	$\xrightarrow[f]{e = \dot{p}} I$
Mechanical	force F	velocity V	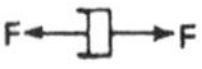 $\xrightarrow[V]{F} R$	 $\xrightarrow[\dot{X}]{F} C$	 $\xrightarrow[V]{\dot{p}} I$
Hydraulic	pressure P	flow rate Q	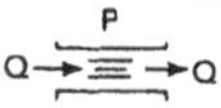 $\xrightarrow[Q]{P} R$	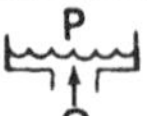 $\xrightarrow[\dot{V}]{P} C$	 $\xrightarrow[Q]{\dot{P}_p} I$
Electrical	voltage e	current i	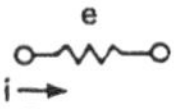 $\xrightarrow[i]{e} R$	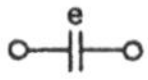 $\xrightarrow[\dot{q}]{e} C$	 $\xrightarrow[i]{\dot{\lambda}} I$
Project (PMSS)	effort e	flow f	 $\xrightarrow[f]{e} R$	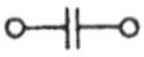 $\xrightarrow[\dot{q}]{e} C$	 $\xrightarrow[f]{\dot{p}} I$
			$\xrightarrow[f = \dot{q}]{e = \dot{p}} Z_i$		

Table 1b. Multiport and source elements.

System	TWO PORTS		MULTIPORTS		SOURCES	
	Trans-former T	Gyrator G	Effort Junction 1	Flow Junction 0	Effort	Flow
Engineering	$e_1 = me_2$ $f_2 = mf_1$	$e_1 = rf_2$ $e_2 = rf_1$	$f_1 = f_n$ $\Sigma e_i = 0$	$e_1 = e_n$ $\Sigma f_i = 0$	S_e	S_f
Project (PMSS)	T $e_1 = me_2$ $f_2 = mf_1$	G $e_1 = rf_2$ $e_2 = rf_1$	A $f_1 = f_n$ $\Sigma e_i = 0$	O $e_1 = e_n$ $\Sigma f_i = 0$	S_e	S_f

In 1959 Henry Paynter of MIT devised a modeling system that was based upon the concept of energy and information flow. Named the Bond Graph method, basic energy dissipation and storage elements were defined and a graphical algebra developed. This technique incorporates the salient features of state space concepts. Properly augmented bond graphs permit direct writing of system state equations.[7] We have found them to be directly transferrable to the PMSS problem addressed in this paper. These thoughts will be developed further after a brief discussion of the family of system elements used in bond graph modeling.

System Modeling Elements

The bond graph system of modeling makes use of three fundamental single-port elements. These are the resistor (R), capacitor(C), and inertia(I). These elements either dissipate (R) or store (C,I) energy.[8] Refer to Table 1a for complete definitions.

The two-port elements transformer (T) and gyrator (G) are also used. Effort (1) and flow (0) junctions along with effort (S_e) and flow (S_f) source elements complete the family.[9] Refer to Table 1b for details.

The relationships between the various modeling elements and system variables are illustrated in Figure 9. The analogous PMSS variables have been added for clarity. These relationships will be discussed in more detail as to how they have been interpreted for PMSS analysis. A summary of the power and energy variables with their project analogies is given in Table 2.

Activity Graph Project Modeling

The simple circuit and bond graph of Figure 10 can be shown to exhibit several properties and characteristics that form direct analogies to the

---

[7] Paynter, Henry M., Analysis and Design of Engineering Systems, Cambridge, MA: MIT Press, 1961.

[8] Karnopp, Dean and Ronald Rosenberg, System Dynamics: A Unified Approach, NY, John Wiley & Sons 1975, pp. 13-65.

[9] Ibid.

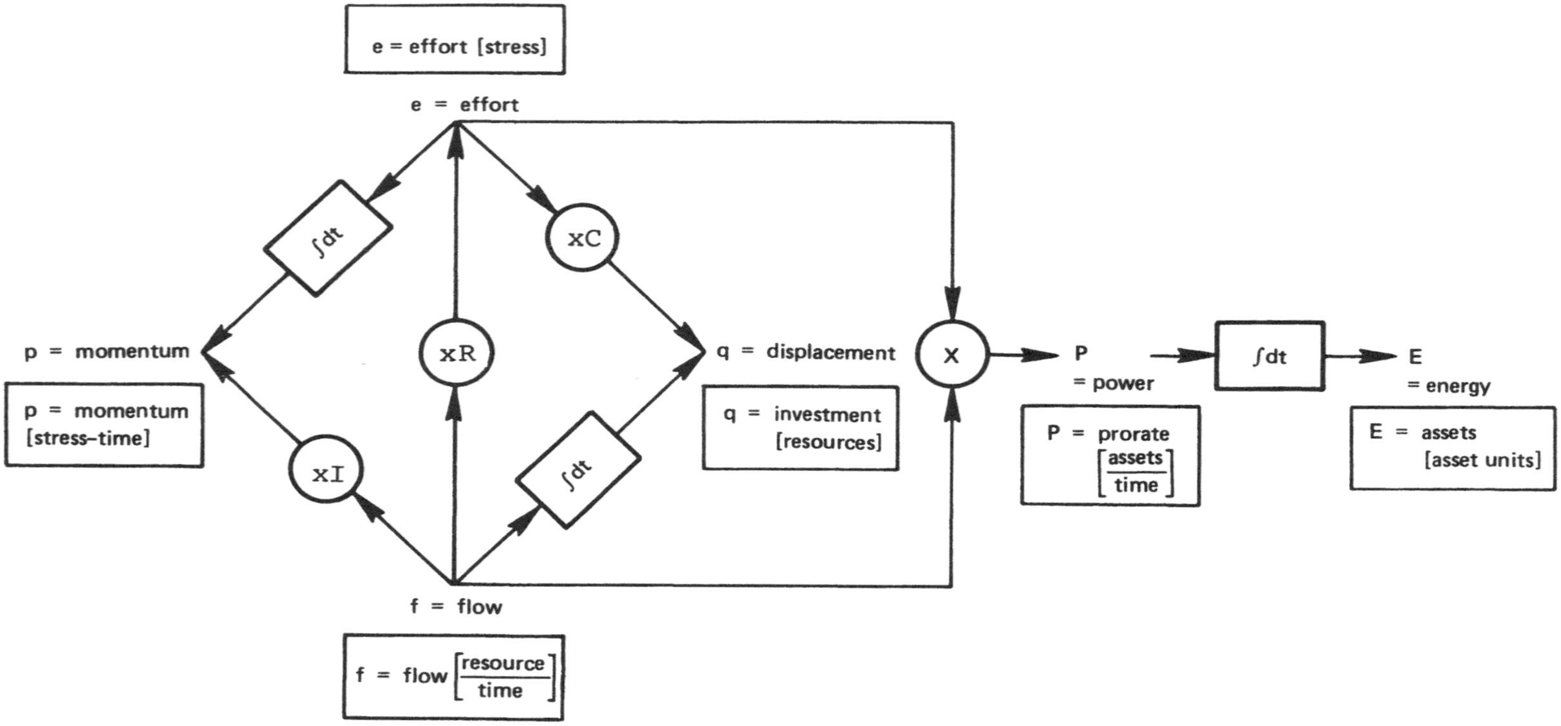

Figure 9. Relationships between modeling elements (R,C,I) and system variables.

Table 2. Power and energy variables.

System	Effort e	Flow f	Momentum $p = \int e\,dt$	Displacement $q = \int f\,dt$	Power $P = e \cdot f$	Energy $E = \int e \cdot f\,dt$
Mechanical	force F [newtons (N)]	velocity V [meters/s]	momentum mV [N-2]	distance X [meters (m)]	power $F \cdot V$ [(N-m)/s]	energy $\int F \cdot V\,dt$ [N-m]
Hydraulic	pressure P [N/m^2]	flow rate Q [m^3/s]	momentum P_p [$(N\text{-}s)/m^2$]	volume V [m^3]	power $P \cdot Q$ [(N-m)/s]	energy $\int P \cdot Q\,dt$ [N-m]
Electrical	voltage e [(N-m)/C]	current i [C/s]	flux linkage λ [V-s]	charge q [A-s]	power $e \cdot i$ [(N-m)/s]	energy $\int e \cdot i\,dt$ [N-m]
Project (PMSS)	effort e $\left[\text{stress} = \frac{t_n}{t_i}\right]$	flow f $\left[\frac{\text{resource}}{\text{time}}\right]$	momentum p [stress-time]	investment q [resources]	prorate $P = e \cdot f$ $\left[\frac{\text{assets}}{\text{time}}\right]$	assets $E = \int e \cdot f\,dt$ [asset units]

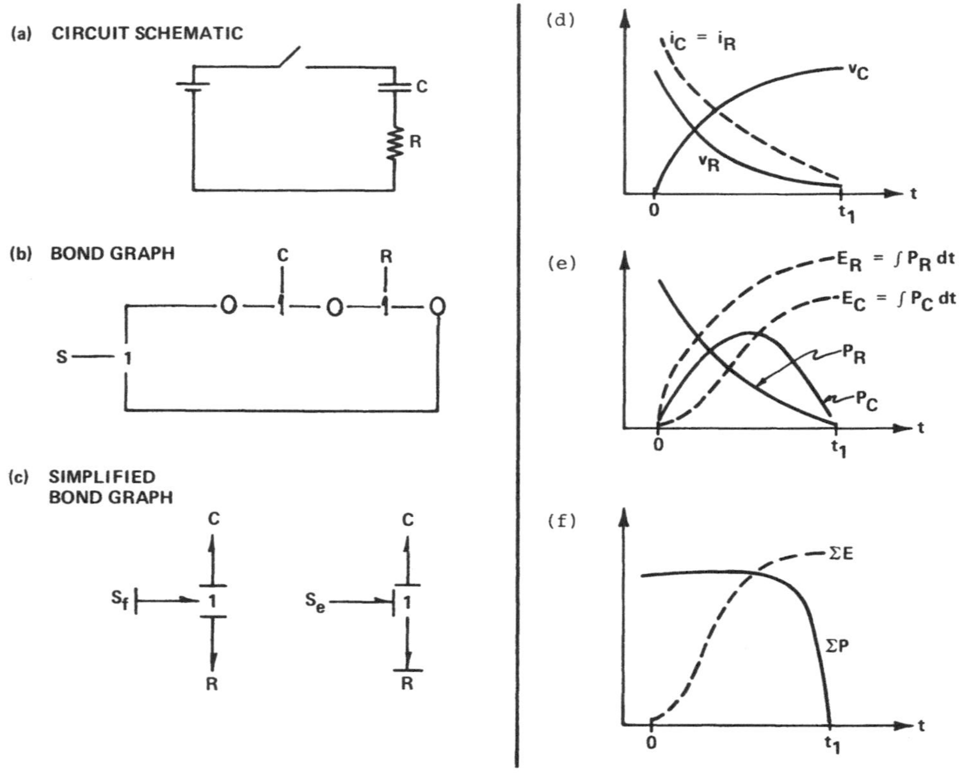

Figure 10. Electrical circuit analysis.

project environment. To illustrate this better, the partially simplified bond graph is repeated in Figure 11(a) (less ground return). Directly beneath it in 11(b) is a simple three-event activity network. These three events correspond with the three "0" flow junctions shown in 11(a). The interpretation is that the circuit element C and the power bonds joining junction numbers 1 and 2 represent activity 12. The dynamic characteristic of this "activity" is that of storing energy, or in PMSS parlance, creating "assets." In similar fashion, the circuit element R and the power bonds joining junction numbers 2 and 3 correspond to activity 23. The characteristic of this "activity" is that of dissipating energy. In the PMSS project situation this is interpreted as expending resources over a period of time without creating any retained assets.

Figure 11(c) illustrates how the "Activity Graph" is defined in relation to both the conventional project activity network and the bond graph. Preserving an activity network type format eases the transition to thinking in activity graph terms. The "A" junction (=activity) is functionally identical to the "1" effort junction in the bond graph. The dynamic characteristics of each activity are modeled by bonding the appropriate R,C and/or I elements to the A junctions.

One of the desirable features of PMSS activity graphs is the ability to analyze the flow of individual resources throughout a project. Thus, the dynamics of individual processes can be modeled more accurately. Assuming this is the case, then Figure 10(d), (e) and (f) are interpreted as showing the levels of effort e, flow f, prorate P, and assets E for the two activities illustrated in Figure 10(c). Accurate modeling of real project dynamics requires that the appropriate R,C, and I elements and their respective values be selected for each activity and resource flow.

An augmented PMSS activity graph is shown in Figure 12. A segment of a conventional activity network appears in 12(a) while an augmented PMSS activity graph is presented in 12(b). The difference in 12(b) is the addition of the activity effort junctions, A, and the dynamic element Z.

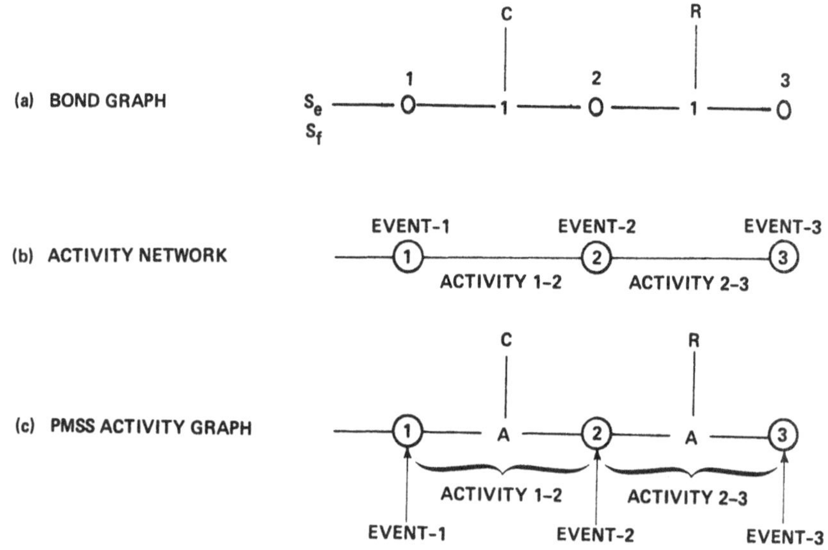

Figure 11. Activity graph project modeling.

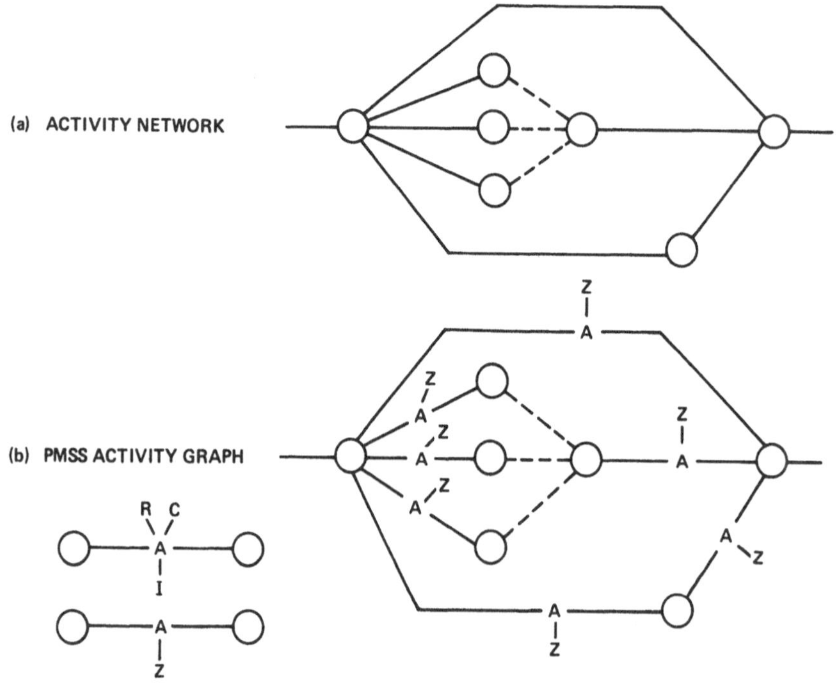

Figure 12. Project activity diagram.

Rather than show R,C., and I separately, the three-component symbol Z (impedance) is used.

Not shown are the subelements of the "event" flow junctions"O". PMSS has three programmable subfunctions imbedded in each of these junctions. These subfunctions are:

1) Threshold detector that verifies when and where a preset level has been realized (e,f,p,q,P or E)

2) Event node that can be set to detect the coincidence of completion of other activities

3) Activity initializer that serves as both a source (S_e or S_f) and trigger for starting subsequent activities.

The addition of these subfunctions greatly enhances the analyst's ability to exercise the PMSS project model. It is this flexibility that allows for experimenting with project dynamics and evaluating various optimization strategies.

Operational Aspects of PMSS

The PMSS approach is currently being implemented in the MACRO-PLANNER project planning and control decision support system. In the following sections we develop an operational view of certain aspects of this support system. Figure 13 summarizes several key concepts and objectives of MACRO-PLANNER.

Modeling Structure

Project modeling with PMSS can involve three or more levels of structure. Going from the summary level down, these are successively called (1) macrostructure/space, (2) microstructure/space, and (3) nanostructure/space. From a conceptual viewpoint, this means that an activity vector in macrospace is represented by the sum of its components as defined in microspace. The same reasoning follows downward for microstructure vectors into nanospace. It is appropriate to think of these structural spaces as being successive hierarchical decompositions of each other.

MACRO-PLANNER characteristics:

- Emphasizes problems critical to large-scale developments
- Data base management systems and development tools
- Analytical tools
- Integration of available program management facilities
- Interactive interface for use by non-computer experts
- Computer graphics and report generation
- Exploratory data analysis
- Natural language and knowledge representation

Figure 13a. MACRO-PLANNER: A macro-engineering project planning and control decision support system.

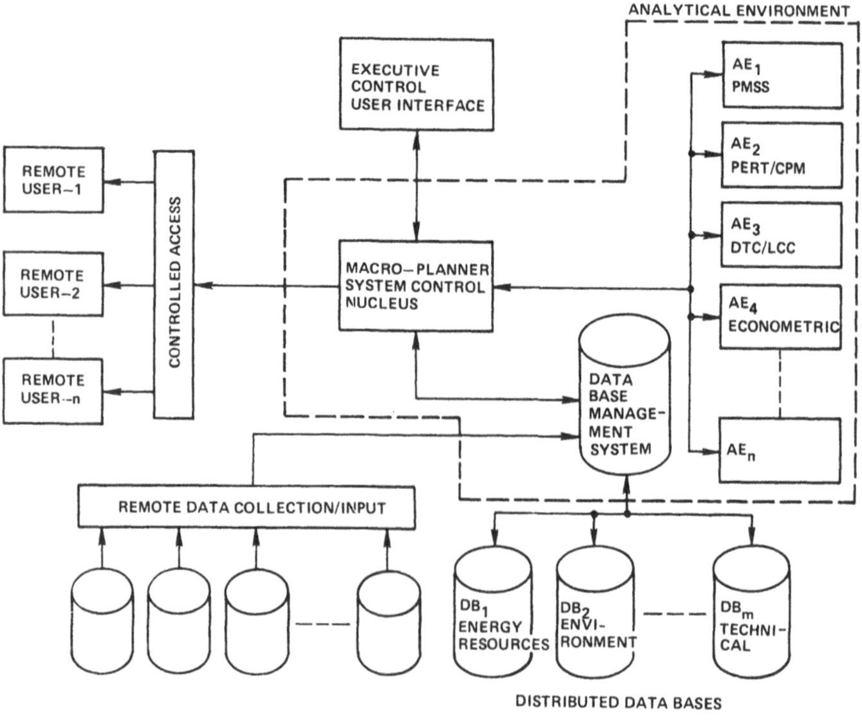

Figure 13b. Macro-engineering program planning and control system architecture.

Data Display

Graphical illustration of state space, where the number of coordinates exceed three, has always been a challenge. Since the dimensions of Project Management State Space will exceed three for all real project models, an early effort was made to develop an effective graphical format, which is essential if one wishes to observe the status of schedules, cost incurred, resource utilization, technical performance, etc., at various points in a project.

The scheme that has been found to be most useful, to date, is illustrated in Figure 14. This represents a "slice" of the project state space at the area of interest. Two state points are shown, indicating the changes over time period Δt. The state points are spread out in two dimensions, each "vector" shown representing the magnitude of one state space coordinate (dimension). A significance to the angular orientation can be defined. Note that an observer in the plane of the figure, looking straight along the line joining the tops of the vectors, would "see" a state point.

The circles shown in Figure 14 represent the desired radial magnitude at time t and $t + \Delta t$. All vector quantities are normalized with respect to them. The actual length of the activity state vectors are superimposed on this display. Vectors not long enough to reach the appropriate circle would indicate activity conditions of behind schedule, cost expenditure under budget, etc. Likewise, vectors extending past the circle indicate conditions of excess.

Additional interactive display formats are being researched for analyzing other aspects of PMSS project models. This interface with the system is critical since it could limit the analysts' effectiveness and efficiency to communicate in the modeling environment. Another graphic display is given in Figure 15, showing a typical "project state space trajectory."

Model Calibration

The PMSS modeling techniques described here require the analyst to synthesize an analytical

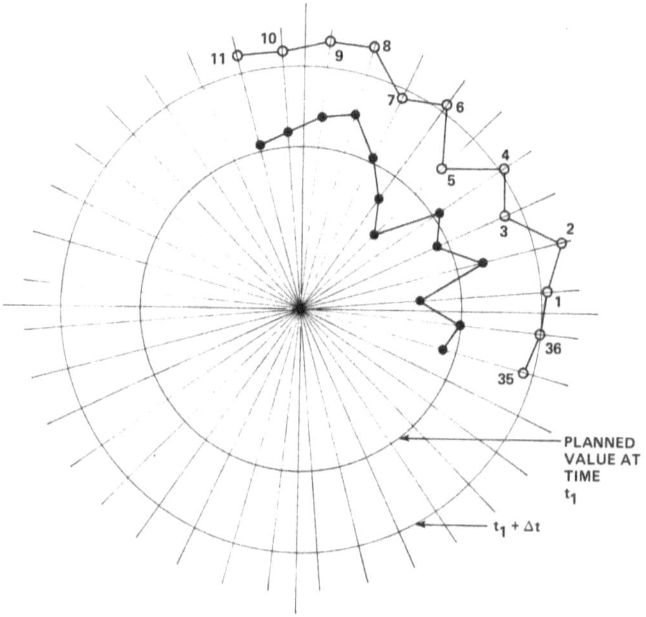

Figure 14. Project trend display.

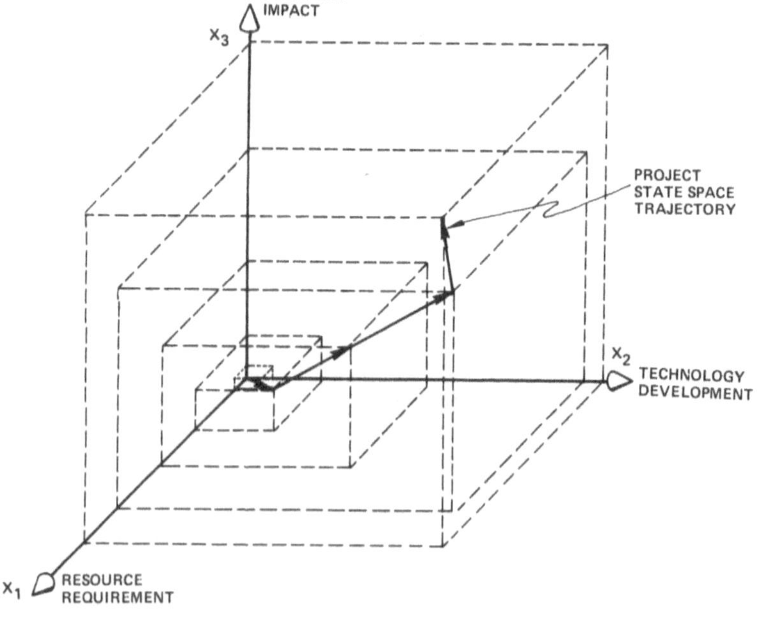

Figure 15. MACRO-PLANNER/project management state space.

representation of the macro-project environment. As with any modeling technique, the result is an abstraction of reality. The degree to which this model accurately represents and anticipates future status of the project depends on not only the authenticity of the technique, but further on the accuracy to which it is calibrated for a particular project situation. With state space models, this process is greatly facilitated when the state variables are observable and measurable.

The state variables that have been defined for the PMSS model are "momentum" (p) and "investment" (q). Momentum has units of "stress-time" and is related to project schedule; investment has units of "resources" and is related to the expenditure of manpower, materials, etc.

The significance of having appropriately defined and observable state variables is that their definition and quantification is sufficient to adequately define the precise state space coordinates of the project trajectory through PMSS. In state space control systems theory, a system is said to be controllable if any initial state $\underline{x}(t_o)$ can be transferred to any final state $\underline{x}(t_f)$ in finite time by some control $\underline{u}(t)$. For the PMSS case, this latter property is interpreted to mean that the state space trajectory (outcome) of a project is predictable (controllable) using PMSS techniques. This capability is of primary interest and value with respect to the planning and control functions of project management.

Knowing the values of the state variables for the actual project over some period of time will allow the analyst to derive what the appropriate values should be for the model's activity impedances. This is essential since the static and dynamic performance characteristics of the model are determined by the system element values and their arrangement in the project model "circuit."

The requirement for adequate validation of any model to determine the inherent accuracy and range of application and thus predict reliability is frequently not thoroughly addressed. This is almost certain to lead to disappointment of the user and discredit to the technique.

This eventuality is being guarded against in the development of the PMSS approach and the MACRO-PLANNER decision support system. Reliable and accurate historical data on several macro-scale projects is being used to verify that the PMSS model can in fact be validated. It is also building confidence in this new modeling technique and its application to the macro-scale project environment. It is further helping to mold the modeling structure as the technique evolves from the concept stage on out to a fully developed and deployable decision support system.

Conclusion

In this discussion, we have presented a framework which characterizes macro-engineering programs and projects from a managerial policy, planning and control perspective. From this framework, we have developed a multi-dimensional and relativistic project state space in which to portray analytic representations of macro-projects. We have derived in this state space a planning and control model, PMSS, which is designed to support the analysis and optimization of certain key project objectives and goals. The planning and control model is implemented and under study in a computer-based macro-engineering project planning and control decision support system called MACRO-PLANNER.

Section B

Values, Psychology and the Decision Process

Abandoned and undisturbed for centuries until its discovery by Yale University's Hiram Bingham in 1911, Machu Picchu justly deserves its title of "Lost City of the Incas." Perched on a saddle between the Machu (Old) and the Huayna (New) peaks, the urban site stretches over some 13 square kilometers. Its construction parallels that of most Inca "pucaras", or fortified city defense units. This important example of a macro-engineering project in the pre-Columbian Western Hemisphere includes a temple and a citadel that were once surrounded by terraced gardens linked by more than 3,000 steps. Eight-thousand feet above sea level and some 2,000 above the rapids of the Urubamba River, it is one of Peru's most impressive ruins.

Overview

Many questions can and should be raised about macro-engineering. What is our scale for measuring or comparing macro-projects? What is the motivation of the macro-engineer? What is the role of systems science (especially in relation to communications, transport and computer technologies) in modern macro-engineering? Finally, how does macro-engineering affect changes in human values as it inevitably will? Ramón Barquín explores these issues in some depth in order to suggest an appropriate conceptual framework for examining important social and cultural impacts.

Ramón Barquín

5. On Large-Scale Undertakings and Human Values

Let me start by saying that I bring no answers, for I have understood my task to be the raising of questions.

Macro-engineering and human values...large-scale undertaking and human values. What do we mean by large-scale undertakings? Clearly, not every large-scale undertaking is a macro-engineering project, though every macro-engineering project is by definition a large-scale undertaking.

Engineering is the science by which the properties of matter and the sources of energy in nature are made useful to man in structures, machines and products. Macro-engineering...large-scale engineering....

Let's talk about scales....

Scales are for measuring...metric...relative dimensions... What must we measure, and what mark must we surpass before an engineering project can be prefixed with a macro? Is the scale in terms of total resources expended...duration of the project...manpower involved? Do we address the magnitude of the effort, or that of the results?

Large-scale, the dictionary tells us, also means: of wide scope. Are we concerned with the size of a structure...the complexity of a project.. the percentage of humanity benefited...or with still other characteristics of an endeavor? Maybe there is a direct relationship between many of these factors. Obviously there often is...

Our most frequent approach -- though not necessarily the best -- seems to be: let's look at cost...a money scale. "About a billion dollars, that's what a macro-engineering project is," a colleague recently volunteered. Are these Hong Kong dollars, U.S. dollars, or Australian dollars? Are they 1920, 1967 or 1978 dollars? What does this really measure? Are we to say that the Empire State Building does not qualify because it came in at a mere $42 million in 1931? In spite of its 60 miles of water pipes, 7 miles of elevator shafts, and the 7 million man-hours it took to complete? Or do we leave out the $650 million World Trade Towers finished in 1973, into whose 201 acres of rentable space, you can march every person in Denmark and stand them a yard apart?

Quite obviously, we can concert currencies and develop equivalency tables across time and space. Yet, even so, we should be very careful when we use a money scale. Can we accurately calculate the cost of building the Great Wall of China, which took place literally over two millenia, using enough material to build 120 copies of the Great Pyramid, and taking a toll of over 300,000 workers' lives during the process?

What's more, the Jubail complex in Saudi Arabia will cost about $20 billion before completion. This clearly qualifies it as a macro-engineering project. But what about efficiencies and inefficiencies? What about management and mismanagement? Would an unnecessary $10 billion cost overrun make Jubail a larger-scale undertaking than it already is?

Wide scope, complexity, orders of magnitude, challenge...in addition to cost. These seem to be the key ingredients. It was macro-engineering at the micro scale of the atom that launched us into the large-scale undertaking of space exploration.

Let us focus on the large-scale engineering of antiquity. Let us ponder the building of the Pyramids of Egypt...or the Grand Canal of China... or the great Mayan cities...or the majestic temples of Borobudur. Should not the scale be relative to the resources and the technology available to a specific society, in each corresponding historical period? Can we find a common scale to evaluate the engineering of Stonehenge and the building of Machu Picchu? Can we compare the development of the Concorde to the construction of the Inca bastion of Sacsayhuaman?

Would we want to?

Let's delve deeper into our subject...what motivates the engineer...the builder? Is it the same set of factors that motivate the macro-engineer...the large-scale builder?

What moved Imhotep -- macro-engineer, physician, and builder, later deified by the Egyptians--to design the Step Pyramid at Saqqara, model for the Great Pyramids of Giza? What made him take the step of constructing one mastaba upon another, breaking with all precedent and tradition? What motivates the macro-engineer?

Better still, really, who is the macro-engineer? Was it Imhotep, the builder; or was it Zoser his Pharaoh? Who conceived the grandiose scheme? Imhotep's name has reached us as the man who broke the molds at Saqqara and showed the way for the colossal works of Khufu, Khafre, and Menkure. However, there is no one name that passes down to us as having been the grand architect of these Great Pyramids. By this point, clearly, the projects had a loftier purpose: a house for all eternity...the proper abode for the immortal. For this, in the case of Khufu alone, it meant the skillful quarrying and masterful arrangement of over two and a quarter million stone blocks, some of them weighing as much as seven and a half tons.

The pharaoh was deemed to be immortal, and yet...might not the actual magnitude of the pyramid tomb that he envisioned and built be an attempt at finding personal reassurance of his immortality? A large-scale structure...a large-scale tomb...a definite attempt to reach out and grasp beyond our mortal world. Are not all large-scale engineering projects to a certain degree a reflection of this same urge?

Thus, we ask again: who is the macro-engineer? Is it the technician hard at work at the drafting table, or solving a series of differential equations? Maybe...but much, much more.

The macro-engineer appears to be the innovator, the grand director, the visionary, the master architect of real-life dreams. But most of all, the true macro-engineer is a man projecting himself to the beyond.

Man builds...man creates...it is in his nature. But when man decides to build in a scale which surpasses by far that which is common, he does so to reach out beyond himself. The justification for his actions may exist, the demand for his work might be there. Yet the grandiose solutions embodied in the macro-projects almost invariably entail this desire to transcend the limiting constraints of our nature.

In many cases,the undertaking might be dictated from above in one individual's search for immortality. In other cases, it might be the concerted action of a people who have decided to leave their mark upon this planet.

Ancient macro-engineering projects seem to have been mostly undertaken precisely by people who left their mark...empire builders in many cases. Was their willingness to engage in large-scale engineering projects related to their success in political, or economic development? Or was it the fact they were politically and economically successful that allowed them the luxury of these macro-endeavors?

Obviously, there are no clear-cut answers, but these two issues appear very related and they seem to reinforce each other. Nation building...empire building...they require commonality of purpose and organization. But, short of war, is there any other activity which achieves these two objectives better than a macro-engineering project? In fact, is there not an important parallel between these two large-scale undertakings in that an unusually large percentage of a people's resources are brought to bear on one sole activity?

It is reputed that at the peak of its construction, one out of every three males in China was working on the Great Wall. I wonder how many of the wars that China fought required a larger proportion of its men in the front lines.

Organization is an inescapable by-product. There is no way to engage in a rational scheme to manage the resources involved in building say, the Mayan city of Tikal, the Roman Aqueducts, or the pyramid complex at Chichen Itza without the development of an organization.

Organization implies some degree of specialization, and the skills developed in resource management and organization for these large-scale undertakings must have been extremely valuable in other national pursuits.

Furthermore, these huge landmarks of antiquity must have also provided a strong sense of national identification. Herodotus, a Greek, first tells us of the great colossi at Giza; but to the man of ancient Egypt, they must have been as much a symbol as the Statue of Liberty is to a New Yorker, or the Eiffel Tower to a Parisian, or to a Cuban the Morro Castle.

Let us come back to resource management and organization. The key element involved is information. Large-scale engineering projects imply the need to record, classify, store, modify and transmit large amounts of information. The demands for information handling capabilities, and man's attempts to fulfill these demands in the context of macro-projects were probably the principal factor leading to the development of an organizational infrastructure. Within these ordered schemes the transmission of information could take place in a more effective and efficient manner.

Communications and computation are critical engineering needs directly related to information. These are two important technologies almost always advanced by any macro-engineering project.

Technology is often defined as "comprising the practical media by which man interacts with his environment to meet his needs." More specifically, a technology is the set of disciplines, methods, techniques and supporting instruments which make up the process by which a tangible or intangible product is elaborated.

In any macro-engineering pursuit many individual technologies are utilized, according to any definition. But in addition, the set of disciplines, methods and techniques which make up the process by which a large-scale engineering project is conceived, designed, planned, financed, implemented and maintained in itself also features many characteristics of a technology proper. Moreover, because its components must interact in a system fashion to produce its output -- the large-scale engineering project -- it resembles a systems technology.

Communications technology, transportation technology, and computer technology are three examples of systems technologies. Today, any large-scale undertaking must almost invariably utilize these as instruments especially geared to handle complex endeavors of wide scope, magnitude and significance. This is the triad of the contemporary macro-engineer. They constitute the infrastructure needed for the macro-projects of our day.

But the computer and communications industries, of course, have also seen their share of large-scale engineering; and the results of many of these projects have had an important impact on mankind, and on human values.

Human values... Is this not also a part of our topic? And we have not addressed it. Or have we?

What is a value? Kurt Baier tells us that it is "an attitude for or against an event or phenomenon..." Kenneth Boulding defines it in economic terms as nothing more than a "preference function." In effect, a value can be just about anything that we prefer in relation to something else. Thus a brief list of common human values might include: one's own pleasure, economic security, friendship, privacy, nationalism, peace, rationality, power, beauty...

We know that the values of a society play a decisive role in human motivation. Values are at the root of almost all social behavior. But should not our main concern with respect to values probably be a dynamic one? Should we not be interested principally with value change? With how values change? When they change...and why?

Value change and technology seem to have a very special relationship, though the mechanics of their interaction are still relatively ill-defined. For example, we have not yet been able to resolve whether values came first -- as in Max Weber's tenet that the Protestant work ethos gave rise to modern technology and capitalism -- or whether it is technology which, in the Marxist sense, determines society's values.

We do know that as a result of technology certain values seem to be acquired or abandoned, others, emphasized or de-emphasized. Alvin Toffler has already said that value impact forecaster is the profession of the future.

Do large-scale engineering projects cause value changes? In many cases, probably yes. Did not the completion of the Transcontinental Railroad and the construction of the Panama Canal have an important impact on some traditional American values? Were not devotion to family life and preference to live in one's home town somewhat altered by enhancing the geographical mobility of the average person? Has not the Manhattan Project, which ushered in the nuclear energy age, had a definite impact on man's peace values, as well as on those related to prudence, and hopefully reverence for life?

In our willingness to do battle with nature and build a massive Grand Coulee or a Boulder Dam.. or construct the Itaipu by diverting the Parana River for four years into a man-made cut...or put a man on the moon...or harness the energy of the atom...are we not in effect underwriting or emphasizing self-reliance, self-fulfillment, rationality, and power?

And, information...the principal commodity of our time...has not its instantaneous transmission through telephone and telegraph, satellites and earth stations, radio and television, to all effects shrunk our planet? And this has surely had an important impact on many of our values.

And what about computers? Their capacity for information-handling has pushed the limits of knowledge much beyond the expectations of prior generations. Random bits of data are usually worthless; but organized information is knowledge, and knowledge is power.

The marriage of computers and communications is allowing us already to develop a common infrastructure for the knowledge-based needs of society. "Think of information as energy," I'm sure you've heard before, and this is exactly right. The future should see each person's access and utilization of information sources greatly enhanced, and with this society's potential for optimizing resource management at the individual level.

We have considered large-scale undertakings, we have considered information, and we have considered human values. As I remarked at the beginning, I bring no answers, I hope I've raised some useful questions.

And, in ending, let me symbolically bring together the past, the present, and the future. For I have the uneasy feeling that through the ages the baton has been handed down from one macro-generation of macro-engineers to another in a continuous fashion: from the Pyramids to the Great Wall, Angkor Wat, Borobudur, and Machu Picchu; from the Taj Mahal to the Brooklyn Bridge to the Empire State Building to Hoover Dam; from the wheel to space travel, from the crank-shaft to computers and jet planes. And today, that baton is in the hands of the men of Itaipu and the TransAmazon Highway, of the planners of the "Chunnel" and the builders of the Jubail. Best that baton rest in steady and secure hands, for the macro-engineers of tomorrow, and mankind with them, will need it as they try to harness the tides and solar power, mine the ocean floor, capture icebergs, or energize the earth from a small black-hole.

Human values will change, there is no doubt. We trust that they will change for the better.

Notes

The quotes from Baier, Boulding and Toffler are from K. Baier and N. Rescher, eds., Values and the Future, The Free Press, New York, 1969, 527 pp.

Information on the Egyptian Pyramids is from A. Fahkry, The Pyramids, The University of Chicago Press, Chicago, IL, 1969, 272 pp.

Khufu, Khafre and Mankure are also known by their Greek names Cheops, Cephren and Mycerinus.

Other data on large-scale projects comes from the Encyclopedia Britannica, and the 1978 Guinness Book of World Records. Definitions are from Webster's New Standard Dictionary.

Overview

The immensity of macro-engineering projects often arouses the emotions of both awe and fear in man. Cynthia Oudejans Harris points out that part of our fear arises because of the feeling that we may not be able to say "no" to macro-projects which technology makes possible but which are not fundamentally desirable. Macro-engineers and planners must develop the ability to say "no" to some macro-projects and become sophisticated about when and why to accept or reject a project. To reduce our fear of macro-projects, we must know that the project's purpose is sound, that the project is appropriate to its purpose, and that we can develop a sense of "ownership" in the project through knowledge about it, participation in its planning, and sharing in its use and development.

Cynthia Oudejans Harris

6. A Psychiatrist Reflects on Macro-Engineering

The title of this symposium, "How Big and Still Beautiful?" tells me that we are beginning to have strange feelings about "bigness." And I suspect that it is because of such feelings that I, as a psychiatrist, was asked to address this gathering. My colleagues and I are, after all, supposed to be specialists in strange feelings.

At the outset let me say that I believe that bigness is here to stay. A little later I shall comment on our gut and emotional reactions to the immensities amongst which we now live. But I wish to point out at the start that we have already intuitively begun to scale down our world: we no longer seek to exaggerate our mastery in our own eyes by creating immense public spaces, as we used to do; we will build no more Pennsylvania Stations, gorgeous though it was. And not only because it's too expensive: we don't need to any more. We have established ourselves as inhabitants of our planet. And in recent years we have found we need more intimate spaces and smaller experiences to balance out our mastery over the sky and over the waters and over the earth....So the big law firms have begun to lower the ceilings in their old and distinguished buildings; hanging plants are everywhere; chamber music has become immensely popular; the brand new airport in Cleveland is filled with numerous spots where just two or three people can sit comfortably together; we have taken to vest-pocket parks; and we have taken to the mini-sport of jogging.

I mention these examples so we won't forget, as we talk about the big things, that we've recently taken to spending time, energy and money on small things, too. And that's very valid.

But back to bigness. We live in a world where the systems on which we depend are growing ever larger and ever more centralized and consolidated, whether or not we wish it. We could not even have come to Houston today were that not so. And, as you know, the 500 largest companies in the United States produce goods worth almost as much as those of the other 400,000 companies put together. Bigness just is the name of the game. We live on a planet which has turned MACRO in our lifetime, we live in a MACRO world.

What does this do to us at the gut level? How can we find our way amidst the immensities surrounding us?

It has been a privilege for me to devote some thought to the subject of immensity and the feelings it evokes in us. I shall then discuss some of the implications for macro-engineering which I see growing out of our response to the immense.

Let me start with something big and beautiful: Niagara Falls. Recently I visited them. I stood on the observation deck and the earth trembled slightly beneath my feet. My face was damp from the mists and I heard the great roar of the falling waters. I was filled with excitement. Then, as I neared the brim of the falls, where the water pours over in a two-foot sheet, I had a split second of fear. Would I be able to resist the urge to throw myself into it? Perhaps in a barrel, as so many before me? And would I, my tiny self, survive the challenge? Naturally, all of these feelings crowded into an instant, and I quickly, as our language says, "returned to my senses." I again became aware of the solid deck beneath my feet and of my own safety before this magnificent spectacle.

I mention these reactions of mine because I believe these emotions are fundamental to our relationship to the immense. This combination of the safety of the tiny spectator with the challenge of immensity, this mix of security and danger, of the sense of our own smallness and at the same time our wish to assert ourselves by confronting the challenge we experience -- all these feelings are parts of the instinctive exhilaration we feel when we confront immensity.

Are we enhanced or overwhelmed? We struggle between these two reactions. This ambivalence, I submit, constitutes the <u>emotional</u> challenge inherent in macro-engineering.

For what about the awe inspired by the immense creations of man?

Historically, human beings were first impressed by immense works of man -- not nature. The seven wonders of the ancient world, listed by Antipater of Sidon in the Second Century before our era, were not natural phenomena: they were all works of man. Like ourselves the ancients were impressed by man-made immensity.

When we feel dwarfed by modern man-made structures, our awe is tinged with the "pale cast of thought." "Who made it? Why did they? Who paid for it?," we wonder. And thus a touch of paranoia is added to our feelings of excitement and of fear.

In addition, as our capacity to create immense things has increased, we have witnessed so many of our solutions turning into our problems: we have seen industrial growth pollute our air and our waters, we have seen agricultural growth leach our soils, macrotankers pollute our seas. We have seen medical progress double the population of India within a generation -- now it stands at 640 million as against 340 million in 1948. And we have found our own powers to do the wrong thing no longer inhibited by the immensity of nature.

In consequence, we have developed a more intense uneasiness about our own power. "How big and still beautiful?" Can it be that Mies van der Rohe, the great modern architect, really <u>was</u> right when he said: "Less is More?" The scariness of "bigness", of the bigness of our projects, and, via them, of the size of our own power has begun to frighten us; we begin to think that maybe small <u>is</u> beautiful, if maybe less <u>is</u> more, and to wonder if big actually <u>is</u> inherently ugly. We wonder, too, if our own <u>power</u> is ugly. We experience a tinge of fear because we are, after all, so small. And a tinge of worry, because, after all, we <u>do</u> have the power, -- and can we be sure that we will use it wisely? And beyond that, we now know that it's all up to us. Neither a god nor nature seems able to control us.

But also we have new freedom.

A hundred and sixty years ago, in 1820, the great German philosopher, Friedrich Hegel, wrote: "The History of the world is none other than the progress of the consciousness of freedom." Still today there is much truth in his statement.

Freedom consists, first, in the availability of choice, of outer choice; and, secondly, from the emotional point of view, freedom means the capability of inner choise; emotionally, freedom is the ability to say "yes" or to say "no" to the given, outer environmental choice. For example, through our institutions our society has given us freedom of speech. That freedom -- which, of course, must be constantly reconquered -- makes available to us the freedom to speak. Then we must choose whether to speak or to be silent.

So now, in Hegel's terms, we have made a piece of progress. We are conscious of possessing a new freedom: we have become outwardly free to build great projects, great tunnels and bridges and air networks, to build new cities in the jungles and deserts, to build great islands in the seas and to begin to create settlements in space.

What can my discipline offer in all this?

In a free society, such as our own, psychiatry is intimately connected with freedom. It is my opinion that all of our best psychiatric activity is devoted to helping people develop the capability of choosing and of acting on, living out of, informed choices. "Mental health," as it is euphemistically called, is nothing more than the ability to make free and responsible choices and to live out of them.

(Of course, freedom is also onerous and difficult -- which is why my profession exists. Sometimes we want very much to have some "Big Brother" relieve us of our freedom, to remove us from the responsibility for our lives.)

So how can we as engineers, planners, informed citizens best live with our new macro-freedom? How can we be more enhanced and less overwhelmed by it?

The first thing we need to do is to develop our own sense of choice. We need to know in our very bones that, no matter how seductive a given project may be, we are at least able to say "no" to it as well as to say "yes" to it.

You all remember Hillary's famous response when asked why he bothered to climb Everest: "Because it is there," he said. We need to embrace the freedom to say: "Well, no, I won't climb it, --even though it is there;" and to say: "Even though this project may be the biggest challenge I will ever face, we will not build it. No, this project should not be built." For remember -- in case you happen to be into challenges -- the biggest challenge may actually be to say "no" to the "biggest challenge."

For, if we must say "yes" because this is our great challenge -- or simply because it is so marvellously exciting -- we are not truly free. To be truly free, we must have "no" as well as "yes", "yes" as well as "no" in our repertoire.

Of course, both our "no's" and our "yes's" -- and also our "maybe's" -- must come out of our own very best knowledge. This is not so simple as we perhaps once imagined. For it may very well be that, even with the aid of our most trenchant data and our most advanced computer techniques, we will still be unable to foresee all that may come about because of an immense project; furthermore, the data and the circumstances have become so complex that it takes more than one mind to encompass the realities being dealt with. Each individual can only comprehend parts of the whole. That is scary, too. Who can really know? How can we, as individuals or as groups, act in a truly realistic and responsible fashion?

Possessing the emotional capability of saying "no" or of saying "yes", is a beginning. Becoming truly sophisticated about when and why to say "no", when and why to say "yes" are skills that we will need constantly to refine.

So far I have been speaking chiefly about us in our capacity as decision makers or in our capacity of informed citizens.

Now I want to turn to the impact of man-made "bigness" on us as consumers, as spectators/ builders/users of the great macro-engineering projects. In and of itself, immensity awakens our awareness of our tinyness -- as I have already mentioned, we easily feel overwhelmed and powerless and a little paranoid when we come upon some immense structure that we didn't even know existed. And if we then imagine that "they" built it with some vaguely thoughtless or perhaps deliberately evil purpose in mind, without consulting us about our needs, wishes, desires -- for sure, this immensity is "bad".

I had some such feeling when recently -- quite by accident, just following a country road along the shore of Lake Erie -- I stumbled upon the nuclear power plant, now being built in Perry, Ohio. There were no signs, no notices, nothing -- only cultivated fields on either side of the dirt road. And suddenly there it was, immense, with a big fence and guards and lots of activity. I hadn't really been quite aware that it existed in my own back yard. I was left with a vague feeling of having encountered something evil. In my head, at least, this place belonged to "them" and not to "us". My guess is that each of you has had at least one similar experience within the past few years.

So again back to Dr. Davidson's guiding question: "How big and still beautiful?" I don't think this question can be answered in terms of meters and kilometers. Beauty, after all, is in the eye of the beholder. What sorts of things do we as consumers, users, spectators of macro-projects need, in order to be able to welcome a macro-project as "Wow! Beautiful!"

There are at least three things we need to be sure about if the big things are not to make us paranoid but rather to be beautiful to us.

The first thing is that we need to be sure that the purpose for which the particular project is developed is a sound one. To lend a big project our support and not simply to resent it for its very size we need to feel that it is a "good thing" and that those who planned and developed it had their heads about them as they planned: that our needs for a clean and healthy natural environment were attended to, that long-term effects were considered, and so forth. As far as I am able to tell, precisely this sort of attention to purpose went into the building of the immense new polders -- 540 square miles of man-made land, so far -- on the bottom of what used to be the Zuider Zee in Holland.

Once we know that the purpose is a good one, the second thing we need to know is that the immensity of the project is appropriate to its purpose and to the available resources. The Dutch wouldn't empty a bay to make a single acre of land; only hundreds make any sense. And on the other hand, Proposition 13 in California has turned several immense projects into white elephants, as for example the gigantic, useless highway interchange -- disconnected from any other highway -- to the north of San José. The immensity of that project became -- and was that foreseeable? -- inappropriate to the available resources.

And the third thing we need to know -- if an immense project is to be big and still beautiful in our eyes -- is that it belongs to "us" -- not to "them". I had no such feeling, peering through the fence at the nuclear plant. It surely didn't belong to "us".

How do we get that feeling of ownership? And what can we do to foster it in others? I want to discuss this process in some detail. Daniel Yankelovich, the economist, has pointed out that one of the human factors contributing to the present surge of inflation is what he calls the "philosophy of entitlement"[1]: we have begun to feel entitled to care by our government when we are unemployed, sick, old or otherwise plagued. I wish to suggest that a "philosophy of ownership" is one of the antidotes to the "philosophy of entitlement" and that we need to pay a great deal of

---

[1] Cited by Robert L. Heilbroner, The New Yorker, August 28, 1978, p. 57.

attention, particularly in the area of macro-engineering.

Not a great deal is known about this important process, but let me share the little that I do know.

Traditionally, we think of "ownership" in financial terms. But to own something financially may or may not help to confer the sense of empowerment, of ownership, to which I refer here. Let me give an example. We talk about "our government" and, although we are weary of paying taxes, we don't feel as if we own the government. But it is "our" government, nonetheless. This is the sense of ownership to which I refer.

As to the process of gaining ownership -- I see three major activities through which we may grow to have a sense of an immense project belonging in some way to us. Doubtless there are other activities as well. The three that I see are: firstly, by becoming knowledgeable about the project; secondly, by having a voice in the decision-making process involved in its planning and consummation; and thirdly, by using and caring for the project.

As to the first way: becoming knowledgeable. Becoming knowledgeable is a primary way in which we gain a sense of ownership about virtually everything and anything. Our knowledge about our scientific fields gives us our capacity to participate in them and hence our sense of ownership of them. And if I know enough about my refrigerator to be able to fix it, I have a much stronger sense of ownership than if I have to call the repairman. Transferring this principle to macro-projects, it is clear that public education about the thinking that went into the planning, about the opportunities and the dangers and the drawbacks involved, is of the essence. And, for this nation of do-it-yourselfers, models and technological data of all sorts should be readily available to everyone. Further, in my own opinion, significant funds should be set aside for just this purpose of informing and educating the public. For, let me stress again, it is not only that data is being passed about: in addition our own sense of ownership, of empowerment, is being nurtured.

The second process by which we may gain a sense of ownership is by having some say in the planning and development of the big projects. Everyone whose life will be intimately affected by any big project should be involved as much as possible in the process by which the plans become reality. At the very least, if your life is going to be affected, you must have an opportunity to have your say about the big structures that are going to have an impact on you. Just having your say will help you feel that in some way the project is yours, that you, too, have some power over the outcome -- even if your wish not to have the new highway involve the destruction of your house cannot be granted. All this, of course, provided that you are heard out in good faith. If you are, having the opportunity to have your say accomplishes a second thing, too: you begin to believe that someone at least gives a damn about what happens to you and to your neighbors. Really to give us consumers a real chance to influence the big plans, or minimally to have our say about what is being done to us and to our environment, is an extremely time-consuming and difficult work; but it is an essential part of emotionally detoxifying immense projects. It is related to the kind of investment of time and energy that a surgeon must make in talking with his patient about the operation he is going to perform. An outstanding example of this kind of investment of effort (and money) was the work done recently by Dalton, Dalton, Newport under the leadership of Engineer Mel Lehr, in soliciting public input into the Baltimore Regional Transportation System Project. Lehr stresses that this was not P.R. work but rather that genuine public involvement in the planning process was a major corporate objective.[2]

A third way in which we come to have a sense of empowerment and ownership is through using a project and, perhaps, even helping to take care of it. We have all had experiences of an enhanced sense of ownership. Our old blue jeans: they have been through so much with us, they really belong to us! And our house is really "ours" -- even if the bank owns most of it -- because we have invested ourselves in fixing it up, turning it into "our" place, putting our own stamp on it. You can all

2 Personal communication.

think of other, perhaps better, examples of how you have come to have an enhanced sense of ownership over something -- perhaps a park, a restaurant, a particular street corner.

On a larger scale, we can say that the biggest macro-engineering project of the past two centuries has been the settlement and development of this country. We have used this country, and in some strange way it has come to belong to us. We and our forefathers explored it, developed it, fought over it, built our homes and our factories and our public buildings on it, and we are here in Houston today trying to help make it a better place -- we have used it and it has become ours. We have a sense of ownership. Even the trash cans admonish us: "Help Keep Our City Clean."

I have been talking, you recall, about three possible activities by which we may come to have a sense of ownership: through getting knowledge, through having a voice, through being a user. And I said that our sense of ownership of macro-engineering projects is important for our sense of ease with them, is important if we are to find them beautiful.

I want to stress that this really is a question of pronouns: if "they" own it and run it, the immense project stands a good chance of being viewed as potentially dangerous at best, as evil at worst; but if "we" have a sense of ownership of the great project, if it "belongs to us," the chances are that it is a "good thing," that it is "beautiful."

Therefore, as consumers, we must ask ourselves: "Does this belong to us? Do we own this at all?" And as planners and engineers and executives we must ever attend to the process of giving ownership of great projects to those who use them or on whose lives they otherwise have impact.

We need to learn a great deal more about the process by which we begin to say "we" and to take ownership. Many issues of trust and honesty, of information and judgment, are obviously involved here. The whole thing is not a simple matter. I have tried to suggest some ways in which our sense of ownership is evoked. I am rather sure that we cannot be manipulated into a fake sense of ownership. It is my own fervent belief that our language is so powerful that we cannot be maneuvered into saying "it's ours" if, deep down within ourselves, we feel it belongs, in actuality, to someone else.

There is a further aspect to all this which I must mention in closing which is of particular importance to macro-engineering projects which are of inter-cultural, international scope. The things I have been saying apply, I am convinced, in our own American culture and substantially in most Western cultures. However, I would not venture to say I know much at all about how an African or an Asian comes to have the kind of sense of ownership which I have been describing -- or even whether such a sense of ownership is appropriate, desirable or possible. Perhaps a quite different group of responses to immensity govern, in another culture. For instance, Akira Kawada, the Japanese psychiatrist, recently wrote about "the unique usage of personal pronouns in Japanese." In the Japanese language the first person is expressed by various pronouns depending on the relationships with persons addressed. For example, the formal word for "I" is "watakushi," but men often use "boku" among friends and "ore" with their wives, intimate friends or subordinates. In other words, the "self" of a Japanese is defined by his relations with others.[3] Even from this limited information about the Japanese language it is clear that "ownership" in Japan must have a very different feel to it than our own. How their use of "we" and "us" differs from ours, I have not had opportunity to discover. But I am sure that Japanese issues around "How big and still beautiful?" are different from our own. And that in still another cultural sphere they would be different again.

When working abroad, it certainly behooves us to sharpen our own awareness to the ways in which foreign perceptions, even of such apparently simple matters as time and space, differ from our own.

In my remarks I have tried to explore our possible inner feelings when we experience

---

[3]Akira Kawada, VOICES: Journal of the American Academy of Psychotherapists, Vol. 14, No. 2, p.42.

Figure 1. Early photo of Henry Wingate, the author's great-grandfather, who was an early settler in the state of Iowa.

ourselves as macro-types; I explored some of the ways we feel when we experience ourselves as micro-types confronting the works of macro-types, and I pointed out some ways in which macro-things may appear less sinister and more beautiful to us.

In closing, I wish to review my main points.

1) Less COULD be more.
Mies van der Rohe was the first to state that "less IS more." As macro-types we need to recall that less MAY be more and so to cultivate the freedom to say "no" as well as "yes" to immense macro-engineering projects.

2) Immensity fills us with awe.
Awe is a mixture of the excitement of challenge with the fear of danger.

3) Awe easily becomes paranoia.
As a micro-type, it was easy for my awe to become paranoia when I stumbled upon the new nuclear power plant now being built near Cleveland, Ohio.

4) It becomes ours when we:
   - know a lot about it
   - have made our own decision about it
   - use it
   - help care for it.

Henry Wingate, my great-grandfather, in his hand the pitchfork with which he helped to make arable and to settle the state of Iowa. You can see that he knew a lot about it, had made his own decision about it, had used it and helped to care for it. And you can see that he felt good about all that.

5) If it's "ours" it's BEAUTIFUL!

6) Different cultures perceive things differently.
In the beginning, only the Europeans wanted to climb the great peaks of Nepal. These mountains provided no challenge to the people who lived near them -- they perceived them differently.

Overview

The principles of social psychiatry, whose application has increased the capacity of both individuals and groups to deal with stress, can help to reduce the adverse social effects of some macro-engineering projects. As explained by Hans R. Huessy, these principles assert that any well-functioning social system provides the following attributes: 1) leadership, 2) social support, 3) opportunities for constructive participation, 4) positive expectations, 5) adequate information, and 6) anticipatory guidance. In many cases, technological advances have removed these attributes from our society. Social engineering on a macro-level is needed to correct the large-scale imbalances which exist. And successful macro-engineering of any kind requires the development of a "macro-morality" in which responsibility for the results of macro-engineering is not shunted aside.

Hans R. Huessy

7. Macro-Engineering and Social Psychiatry

Our topic, "How Big and Still Beautiful," was meant to question the effect macro-engineering projects have had on their beneficiaries, human beings. Unfortunately, the answers provided in our discussions were not wholly adequate.

Perhaps we should look at some early macro-engineering projects and see what effect they have had on people. One of the first big apartment houses built in Marseilles had a playground on its roof. The building made the architect very famous, even though the playground invited disaster, since every pervert who wanted access to children could commit crimes unobserved by any other adult. A garden apartment on the roof solved the problem. We have always had difficulties in our large apartment houses with the hallways, which are havens for crime. I have never understood why hallways are built without any windows from the adjoining apartments. Anyone in a hallway would have to worry about being observed--an unattractive nuisance, certainly, in the eyes of the criminal. Habitat at the Montreal World's Fair came to grips with this; hallways were only on every fourth floor, exposed. And the space they occupied could be used well by the occupants.

We don't have to look too far back to see even large-scale mistakes made in our attempts to provide efficiency and good living through macro-engineering. This country built huge subsidized, low-cost housing projects for multi-problem families. We have had to abandon many of these projects because, to our surprise, hundreds of multi-problem families housed in one area didn't stop having problems. From all we know of how groups work, we should have been able to predict this, and should instead have encouraged subsidized apartment programs for multi-problem families placed among non-problem families.

The construction of large planned communities like Reston, Virginia, and Columbia, Maryland, outside Washington, D.C., and some new towns built in England, also should teach us some lessons. Much effort was spent in Reston and Columbia to provide adequate social organization, but it did not work out. One reason was that the population attracted to these cities at the beginning was transient, with an average stay of one to two years. Second, most of the mothers worked, so the children came home to uninhabited neighborhoods. And third, no older generation was present. The English towns provided their own industries for employment, giving greater population stability, but also suffered from having an absent older generation. Consequently, these big projects, very carefully engineered to accommodate workers and their families, lacked social structure. One curious aspect found among these inhabitants was an overuse of medical facilities. Apparently the doctor's waiting room, not the neighborhood or work room, was a good place to meet people, even though an expensive one.

How can we house or provide work space for a great number of people and still sustain a decent way of life for them? Dr. Cynthia Harris emphasized our need to develop a sense of ownership in the projects' beneficiaries, and provide the public with adequate information about the projects. She suggested too, that we should allow them to feel that they had participated in the decision-making about the project.

This leads me to think that perhaps all designers of macro-engineering projects should keep in mind the principles of social psychiatry, whose application has increased the capacity of both individuals and groups to deal with stress. These principles state that any well-functioning social system provides the following attributes: 1) leadership, 2) social support, 3) opportunities for constructive participation, 4) positive expectations, 5) adequate information and 6) anticipatory guidance.

Let me give brief examples of these attributes. We know that good leadership in a military unit or an institution makes for healthy functioning. When leadership is poor, even basic functioning deteriorates. When a military unit has an unusually high psychiatric casualty rate, poor leadership is the most common cause. The second attribute is social support. We send a Red Cross wagon out to disasters not because the people need coffee and doughnuts, but because the presence of the Red Cross is an indicator that the larger world cares and is supportive.

A very obvious example of the third attribute-opportunities for constructive participation -- comes from an experience at the end of World War II. The Japanese were using suicide planes and when one of these planes hit a ship, its bomb didn't always go off. The plane's gasoline, however, might be spilled around an anti-aircraft gun on the ship, so that the gun crew could not fire the weapon. Due to some Navy regulation, the crew had to stay at the gun, even though it could not be used; and the psychiatric casualty rate would be as high as 60% to 70% because planes were coming at them and they were helpless. There is burgeoning literature showing that helplessness in many forms has deleterious effects on human functioning. (The regulation was soon changed.)

The importance of the fourth attribute, positive expectations, is well-known in schools, of course, and I commend to you the literature on the impact of "labelling." The fifth attribute, providing adequate information, is self-explanatory. Finally, with the sixth attribute, anticipatory guidance, we prepare people for stressful situations which we know they will encounter. A trial run increases their ability to cope.

Buildings, bridges, tunnels and roads are not the only kind of big projects. Creating social organizations can also be a big project. We might use William James' proposal for a moral equivalent for war, or my father's ideas on planetary service as an example. In essence, all young people would give two years of service to their country, to do the things which need doing but will never be economically feasible or offer careers. This could mean our providing genuine caring to the chronic physically- or mentally-handicapped, or a plentiful supply of teacher-companions to children from deprived backgrounds. They could clean up the environment or engage in military service, the Peace Corps or any other similar activity. All these tasks need to be done, but do not offer lifetime careers. Intensive caring of the disabled requires people. But it is so exhausting that most of us cannot do it except for a time-limited period, six months to two years. There is no career in being a Big Brother to a needy adolescent; there is no career in enriching the lives of the chronically handicapped, but it needs doing. The Peace Corps and VISTA are small-scale attempts to offer opportunities for service, and the Civilian Conservation Corps of the 30's served some of these same purposes.

We tend today to assess the needs of a group, and then try to meet that need by paying some professionals to do so. Such programs tend to have little stability through time. We would recommend more "complementary" programming in which two groups meet each others' needs. For instance, the Foster Grandparents' program meets the needs of both the children and the grandparents. Likewise, while the Peace Corps met a need in underdeveloped countries, it also met the needs of young Americans. The role of the professional is to supply the framework only. I believe complementary programming could be developed on a very big scale.

Universal Service would bring together young people from different classes, and backgrounds, and help them feel that they had a personal stake in their country. The youth who feels useless and unwanted would experience being needed. The appeal revolutionary activity has occasionally had for these young people would be diminished and capitalism, the goose that laid the golden egg in this country, would gain more time to solve its human problem.

Along the same lines, if we can build space platforms to generate electricity, we should be able to link a big social project with a big physical project. We need to build adequate housing for low-income Americans. There may not be much money to be made in such an undertaking, but it may be crucial to our survival. Could we use some of our unemployed workers for this?

We need macro-engineering in social engineering, but we must not let ourselves be dazzled by either the size or innovativeness of a project. In programs for humans it is often best to start small. The recent change-over to modern math in public education is a good example. Social engineers in ivory towers decided that by teaching arithmetic as though it were algebra, children would be better prepared to deal with abstract mathematical concepts they would encounter later. Up to that time, one of the few school subjects in which poor children did well was arithmetic. By turning the arithmetic into early algebra, we guaranteed that the poor no longer could have even this achievement. Instead of trying the new math in a few school systems to see what would happen, it was pushed onto the whole country because it seemed so reasonable. "Reasonable" or Cartesian thinking is excellent in the science of things, but it does not work when applied to human situations. Thinking should not precede experience, but must follow it; and however reasonable something may look, we will not know how it affects human beings until we try it. Rational thinking cannot deal with values or morals, which are at the heart of so many human dilemmas.

Another recent example of a disaster in social macro-engineering is the American love-affair with psycho-analysis, a theory based on non-verifiable hypotheses, which implied both the prevention and cure of psychiatric disorder and problems of living. A society priding itself for its rational thinking embraced a completely irrational enterprise. The theory speculates about causes of psychological disorder. Treatment is then based on the same theory and aimed at the supposed cause. Medicine does not treat causes. Knowing the cause of an illness makes possible its prevention but not necessarily its treatment. Knowing the treatment of an illness does not help one to prevent it. A treatment should be judged by its efficacy and not by its theoretical fit. If we were to treat a smoker for cancer of the lung, treating his smoking would be an inadequate intervention. If I were drunk, fell into an ice-cold river and developed pneumonia, the mental health professionals would propose heating the river. I would want a shot of penicillin. With this kind of intellectualization so prevalent in mental health, it is no wonder that no one ever realized that Freud's description of how children develop was unrelated to how children were actually raised in his day. In Europe, all homes but those in the very lowest classes had maids, and perhaps even nannies, who spent much more time with the children than the mothers. And although these people played very important roles in children's development, often being lifetime members of the family, they are not even mentioned in his psychoanalytic theories. We are only now beginning our recovery from this disaster of social engineering.

Big social projects are needed to offset the effects of the technological revolution. All technological progress destroys an existing social group and increases the space in which we operate. When the small town utility is absorbed in a large, regional utility, the group which ran it disbands and the area of service for that utility becomes much larger. Progress, per se, is not bad; but we must actively deal with the destruction of existing social groups by seeing to it that we make possible the creation of new social groups. In the past, a social group primarily existed in a given space. But with technological progress, the ever-increasing spaces in which we move may no longer provide workable boundaries for groups and, therefore, social groupings of the future may have to be time- or goal-based. That is, young people may dedicate several years to the service of their country in groups, or others may work hard together on a particular project for a short period of time. In the past, most social groupings were based on a geographical proximity. Now we need new groups which operate in larger spheres, held together by a common task, some of them time-limited. My father felt that five to seven years was the optimum span for any fixed social constellation in the arena of work. We must learn about the laws of other types of groups. Periods of shared service, as in war, can produce intense and lasting comradeship. Large national organiations provide lifetime friendships for others.

The macro-engineering of buildings and roads is impressive, but easier than the macro-engineering applied to human society. Here follow some dilemmas of our current human engineering: We praise the family, and yet have a welfare system with many regulations that tend to destroy families. We have unemployment, but will spend a quarter of a million dollars to train a handicapped person to take a job away from an able person. We rehabilitate a chronic mental patient to the point where he can hold a job, but as soon as he achieves this economic success, we remove his medical benefits and social support system without which he cannot function. He can seldom earn enough to finance his required medical and social support system. We moan about high medical costs, but freely pay for expensive, unproven treatments and will not pay for much cheaper forms of treatment.

Finally, we must address the lynchpin in successful macro-engineering of any kind--the problem of responsibility. Our moral system defines individual sins and responsibilities, but in this era of macro-engineering, individual sin is not very important compared to collective and corporate sinning. For example, a large group of moral people jointly might do something immoral and no single person can be held responsible. The most crucial immoral acts today are committed by large organizations. Who will do the penance? Today we sin not as individuals but as members of large groups. We have no mechanism yet for a "macro-morality" to go with our big projects. This we must soon address, since we cannot afford to turn our backs on it. Macro-engineering projects have moral aspects. Their neglect has serious negative social consequences. How can a corporation, an artificial person, judged on a short-term financial balance sheet, be moral? Some recent court decisions are increasing the liabilities and responsibilities of corporate directors. Perhaps the legal concept of a corporation needs to be revised to something more akin to a partnership with greater personal responsibility for the consequences of corporate action.

Frank Davidson, our Chairman, and Dr. Cynthia Harris, a noted psychiatrist, both devoted a part of their lives to making the concept of voluntary service a reality. Their undertaking was terminated by the advent of World War II. That termination did not negate the validity of the attempt, but delayed its execution. They began a social macro-engineering project on a small scale, as is proper. This country seems to be reconsidering the value of such efforts. A recent report by the Department of Defense comes to the conclusion that a National Service Program as suggested earlier may be the only method for fairly supplying the necessary manpower for a successful military organization. Perhaps it is time now to think about beginning again. A proper perspective on our priorities, and the application of sound macro-engineering to our social problems could provide us a safe and better future.

Bibliography

Hausman, William, and Rioch, David; "Military Psychiatry: A Prototype of Social and Preventive Psychiatry in the United States." Arch. of Gen. Psychiatry, Vol. 16, June 1967, p.727

Huessy, Hans R. Caring, the Stepchild of Our Medical Care System. In Press.

James, William, "Moral Equivalent of War." Memories and Studies, New York, Longman's Green & Co., 1911

Preiss, Jack J. Camp William James. ARGO Books, Norwich, VT 1978.

Rosenstock-Huessy, Eugen, Planetary Service. ARGO Books, Norwich, VT. 1978

Rosenstock-Huessy, Eugen, Soziologie, Kohlhammer, Stuttgart, Germany. 1956 and 1958.

Dept. of Defense, "America's Volunteers", OASD, (MRA&L) Rm. 3E-773, Pentagon, Wash., DC 20301.

Overview

For those who prefer a long-range point of view, a "macro-chronic" timespan, the effect of technological efforts on future generations is a critical element of assessment. We need a macro-decision process which allows us to consider the desirability of macro-projects as well as their viability. Several contributors to this volume have addressed a number of the questions which arise in connection with the design and testing of such a macro-decision process. The development of a macro-decision system is itself a "macro" project which could have a notable impact on society. Large-scale enterprises have as much potential for benefit as for harm: it is up to us--to the public and the professions--to develop the means of selecting, using and controlling them in our best interests.

8. How Big and Still Beautiful? A Reprise

"Beauty is truth, truth beauty, --
that is all
Ye know on earth, and all ye need to know."
--John Keats: Ode on a Grecian Urn

....And on the pedestal these words appear:
"My name is Ozymandias, king of kings:
Look on my works, ye Mighty, and despair!"
Nothing beside remains. Round the decay
Of that colossal wreck, boundless and bare
The lone and level sands stretch far away.
Percy Bysshe Shelley: Ozymandias

In a bygone age, Keats found beauty, truth, and immortality in the enduring work of a long-dead and forgotten craftsman; Shelley found irony in the ostentatious self-puffery of an ancient macro-engineer. Today we ask again the poets' questions: what is beauty, that we may know it and create it; what is worth, that we may leave it to posterity? Are we potters and artisans? Or are we Ozymandias? How can we know?

To those who have adopted a long-range point of view -- call it a macro-chronic timespan, extending beyond their own lifetimes -- the importance of these questions is manifest. Today's technologies are as capable of producing destruction as they are hopeful of producing good works. What is good in the short run may be disastrous in the long. To achieve the admiration of the second, fifth, or tenth generation down the line may require sacrifice in the present. To avoid the helpless curses of a posterity without options, we may have to forego major projects, programs or achievements in the present.

To those of the macro-chronic persuasion, these thoughts give rise to two operational questions:

1) How can we measure the long-range effects of currently-proposed large-scale activities?

2) If measurement were possible, how could we control or stop the activities thought damaging to the future?

Not all of the thinkers represented here -- and certainly not all of our engineers and voters and financiers -- are believers in the macro-chronic point of view. As a species, we are only a Darwinian eye-blink away from those common ancestors whose critical time-frame was measured in the hours required to procure food, or the days required to find shelter from the onset of winter. It cannot be surprising that most of us still think of project costs in terms of what we pay from our own pockets, and program benefits in terms of our own livelihoods. Especially in the industrialized West, we are a micro-chronic society, whose interest and attention are confined to its own life span.

To the micro-chronic, the poets' questions are meaningless. Beauty is what pleases me now, and worth is wealth, they say.

Ozymandias, as a micro-chronic thinker, probably died happy.

The Problem

George Kozmetsky, Frank Davidson, William Jones and James McHale have outlined the problem. In its applications of large-scale technology, says Jones, society is operating outside the limits of its own control systems. Although some major projects are easily identified, so that long-range impacts and the need for control can at least be considered, other activities of equal potential for harm are not seen as macro-systematic changes until long after the fact. Macro-systems are capable of self-generated growth, without consideration of the dislocations they may cause in the future.

Kozmetsky points out that the managerial planning and control model which makes large-scale projects and programs feasible is not appropriate for their evaluation. We have the ability to organize,finance, and support huge undertakings, but the same training which gives us that ability prevents us from seeing the activities holistically. Traditional managerial models obscure the purposes, interactions, and long-range dynamics of the activities they have made possible. Macro-programs become institutions in themselves, but without built-in control mechanisms for self-examination, evaluation, or alteration of their program goals. We simply have no models to use in evaluating macro-programs.

Davidson points out the essential role of purpose in the structure and behavior of systems: without purpose, there is no system, and without system there is no possibility of control. A fundamental prerequisite for evaluation is the ability to "bring things together", so that purpose may be inferred, made explicit, and exposed to public debate.

The world macro-problem has been aptly defined by Willis Harman, cited here by Kozmetsky:

> ...perfectly reasonable micro-decisions currently are adding up to largely unsatisfactory macro-decisions.

McHale's paper provides a shocking case example of the world macro-problem in action. "Agribusiness" -- the collective noun referring to all large-scale, capital-intensive agriculture -- is a macro-system whose purpose appears to be maximization of current return on capital in food production, through concentration of landholdings, mechanization, and fertilization. The system is driven by the rational decisions of well-trained corporate executives, and its first-order effect is the production of high yields per acre under cultivation. Its longer-range effects, seen with a macro-chronic viewpoint, are the demise of the family farm and the exhaustion of agriculture's most basic resource -- the land itself. Not noted by McHale, but pertinent where "agribusiness" is applied in formerly peasant-farming economies, is large-scale unemployment and migration of the peasants to urban ghettos, when farming shifts from labor to capital intensity.

In agribusiness, farm equipment costing hundreds of thousands of dollars is operated by technicians, not by farmers. Economics dictates that the equipment keep running; stopping it in order to examine the soil is simply not justified. That the land is becoming a sterile medium in which chemical fertilizers may be converted to crops is not important to the macro-farmer or his industrial suppliers; it is rational for them to assume a constant supply of the fertilizers on which they are more and more dependent. The sum of their rational decisions -- all based on the traditional management planning and control model underlying corporate operations -- could produce a macro-problem for all food consumers, should the supply of fertilizers fail. But the managers of agribusiness have not the models, the time-frame, or the language with which to hear McHale's warnings.

The Impact of Immensity

Davidson, Cynthia Harris, and Ramon Barquín have explored, from varied points of view, our human reactions in the face of immensity. Here we all have a common starting point, in our own feelings as we experience the macro-structures and phenomena of nature. The Grand Canyon, the Himalayas, Niagara Falls -- all fill their witnesses with feelings of awe, described by Harris as a mixture of safety and security on the one hand, with danger and excitement on the other. Natural phenomena, too, produce these feelings, as attested by those who have lived through hurricanes or seen the coal-black disc of the sun in total eclipse.

To Harris's ingredients of awe, I should add another -- the feeling of individual insignificance in comparison with Nature's scale and power. Erwin Schell, founder of the department which became MIT's Sloan School of Management, summed up that feeling with a story of a terrified three-year-old's first visit to an ocean beach. When his baffled parents finally quieted the boy and asked why he was afraid, he sobbed, "Oh, Mommy -- ocean so big. Billy so little." We are all puny and powerless, beside nature's monuments. We, and implicitly all of mankind, are simply -- insignificant.

Harris notes that the emotion of fear is an ingredient of awe. She says, further, that some people may be exhilarated and enhanced by their confrontations with immensity, while others may be overwhelmed. I suggest that the fear might be directly related to the individual's own sense of values -- that an ethno-centric orientation, seeing the individual, the species, and Nature as parts of one grand scheme, might contribute to exhilaration, while an ego-centric orientation, seeing the self as the center and purpose of life, might lead to feelings of some inadequacy and meaninglessness when confronted with the scale of natural grandeur.

The ethno-centric value system would also be consistent with a macro-chronic time frame, and the ego-centric with the micro-chronic. And, as value systems are generally thought to be culture-dependent, we might expect a difference in reaction to immensity in peoples of differing origins and backgrounds. It is difficult to imagine, for example, a member of the Hopi Indian tribe being overwhelmed by the spectacle of the Grand Canyon. To the American Indian as to followers of many Eastern philosophies, as I understand them, humanity and nature are parts of an integral whole. They belong together, and where there is a sense of unity and eternity, there is no room for fear of the immense.

Human values are at the forefront of Barquín's tour through the archaeological remains of ancient macro-engineering projects, as well. He notes that the values of their builders--particularly their inferred dreams of immortality -- have determined what has been done in the past. Further, the accomplishments of the past have influenced the values of today, and we may surmise today's projects will leave a mark on the generations to come. For good or ill, what we undertake will build the world in which posterity develops its own sense of "good", "bad", and "big."

Barquín's speculation on the relationship between macro-engineering and the search for immortality is best illustrated, for me, in Harris's delightful story of Mies van der Rohe. The master architect, surveying his finished work, nods his head and muses, "Well, they will know we were here." It might have been Imhotep. The master builder leaves enduring evidence of the accomplishment, as proof that men were there. Who "they" are is immaterial. The designers and laborers of Hadrian's Wall, Stonehenge, the Tower of London, the Roman Forum, and the Rhineland castles had no notion of American tourists, but their works give me a sense of pride, awe, and continuity with the past. Whether the constructions were built for war or worship, commerce or comic relief, does not matter to me. They were built by people, and they remain significant. I am a person, and I share the builders' satisfaction in saying, "They will know we were here." There is meaning in being human.

Barquín, Kozmetsky, and Larry Meador have highlighted the essential role of organization and information handling in the accomplishment of large-scale activities, now as in the past. That organization itself is a macro-activity with its own set of deleterious side effects is recognized in the literatures of Organization Behaviour and Sociology but only rarely in the field of engineering. Let me follow Cynthia Harris's Niagara Falls anecdote with one of my own, to illustrate.

I recently had the chance to re-visit the Falls, with family and a visiting Polish graduate student. Arriving in Buffalo late on a Thursday night, we found our friends there extremely skeptical about the Polish guest's chances of admission to Canada, where the vista of Niagara is far more impressive than the view from the American side of the river. Even if we succeeded in getting Mira into Canada, they said, we would find it impossible for her to return to the United States; her visa was clearly marked "one entry only," and the U.S. Immigration and Naturalization Service was known to interpret that phrase strictly. One Buffalo friend had recently spent three days trying to bail out a visiting Japanese businessman who had casually walked into Canada to see the Falls; that visitor had to travel to Toronto to procure a second American visa, for re-entry to the country.

Hopefully, I took Mira to the Canadian consulate in Buffalo. After a long wait in line, we learned it takes two full working days to procure a visa for a Polish passport: Mira's could be ready by the following Thursday morning. Since she would be back in Gdynia on the following Thursday, it would be impossible for her to see the Falls from the Canadian side. There was nobody in the office who could talk to her directly, on the spot. There was no superior officer we could talk to directly, or reach by telephone. There was no channel for handling exceptions. No, nothing could be done. A call to the U.S. Border Patrol indicated nothing could be done about the problem of re-entry on a one-time visa, either. Period.

If Mira is typical, the Poles are a very persistent people, who take their bureaucracy with a grain of Baltic salt. Getting into Canada became a challenge. Mira spent hours practicing answers to the possible questions border guards might ask, trying to assume a harmless American accent. No luck. She practiced crying, just in case. On Saturday morning, we set out for the bridge.

Stopping first at the American side to get a feel for the re-entry problem, we found an officer who listened sympathetically. "See me on your way back, and don't let anybody take away this slip of paper stapled to your passport," he said. We walked on to the Canadian entry point. "It can't be done," said the guard, and, "highly irregular," said his boss. But a half-hour later, Mira had an official Canadian visa.

The Falls are immense, and we spent the whole day getting to know them, from all possible angles and elevations. Less thrilling, but impressive in its own right, is the Niagara Power Project -- the gargantuan but unobtrusive macro-engineering project which provides electricity to the Northeast while respecting the grandeur of the Falls. We all came away from the day with a dual impression -- the sense of awe and wonder at the Falls, and the sense of conquest. Nobody conquers the Falls, of course-not even the barrel-riders -- but we had beaten the organizational system which surrounds them. To Mira, that demonstration that the individual can defeat an impersonal bureaucracy was just as important a thrill as the sight of the torrent itself.

Criteria for Judging Big Things "Good"

Several of our writers here have enumerated criteria by which proposed macro-activities might be evaluated. Jones suggested measuring the demands of the activity on our natural resource base, the number of people involved in its construction and use, and the timespan in which its side effects would require safeguards. Harris suggests that the purpose of the activity must be worthy, and appropriate to the scale of resource use, and further that the activity have the property of "belonging" to all of us; Davidson also stresses the careful examination of purpose. Barquín adds the need for specific examination of the activity's impact on human values.

Several writers have also referred to the criteria for defining something as "big", without specific attention to its goodness or badness. Among these criteria are sheer bulk; complexity in relation to what has gone before; technological challenge; cost, relative to the capacity of the society contemplating the project; duration of the design and construction; duration of the activity's utility, and the cost of removing it if it should become annoying or dangerous to the citizens of the future.

Implicit in all these thoughts is the notion of control. We are technologically capable of destroying the future. We need criteria of "bigness" in order to tell <u>what</u> to control, and criteria of "goodness" to see <u>how</u> to control. Many speakers have suggested that in our society macro-projects and programs are possible only with government support, and therefore, they are easy to isolate. Jones disagrees, noting the incremental nature of some developments whose impact is monumental. McHale would join Jones: agribusiness is a perfect example. Harman's problem statement, that the grand sum of perfectly rational micro-decisions can produce a macro-problem, remains as our dilemma.

The evaluative criteria listed appear to fall into two broad classes. Utility, economy, and practicality seem to cluster together to produce an overall criterion of project or program viability. Safety, audacity, immortality, and humanity seem to add up to an overall criterion of its desirability. A macro-activity may possess either attribute without the other; only if it possesses both should it go forward.

And this is the heart of the dilemma. The management planning and control model used by business and government leaders is a magnificent machine for evaluating the viability of a proposal. We have no comparable machine for evaluating its desirability. Without such a mechanism, all our talk of criteria for judging "goodness" is a waste of time.

The Macro-Decision Process

The management planning and control model does not simply ignore the desirability of a proposal. Rather, it literally devalues the long-range future where most of the factors making up desirability (as defined here) are focussed. The devaluation is accomplished through the use of the most basic tools of financial management -- cost/benefit calculations and discounted cash flow analysis (DCF).

Discounted Cash Flow is simply the process of comparing what you will get from a dollar's investment in the project, against what you would gain from putting the same dollar in a savings account. The technique asks, "What is the value, right now, of the returns we will get from this investment in, say 25 years?" It turns out that the present value of a dollar that far away in the future is very small, at any realistic interest rate. Effectively, the long-range future is discounted out of existence. Only the short-run returns of the possible investment count as benefits to the financier. And managerial vision is constrained to the time horizon of its tools.

In teaching the skills of strategic decision-making to the managers of the future, instructors of Business Policy focus on four major questions which must be considered in the formulation of strategy:

1) What might we do? (What alternatives have we?)

2) What can we do? (What resources, strengths, and capabilities can we bring to bear on the problem?)

3) What do we want to do? (What is consistent with the values and desires of the organization's owners and managers?)

4) What should we do? (What are our obligations and responsibilities to society as a whole?)[1]

The same four questions are applied to the evaluation of all macro-proposals. The problem we face here is that the questions are asked and answered by the same people -- by those with the ideas, the resources, and the power to decide. The first question -- what we might do -- is answered by those individuals of creative genius who come up with the ideas. They are the visionaries, the dreamers who say, "This is possible."

The second question -- what can we do -- is answered by the financiers and managers (in business and/or in government) who can command the necessary resources. They are the technologists, the organizers, and the movers who say, "This is viable."

And, in our current macro-decision system, that is all that is required to get a project started. The third and fourth questions -- adding up to what is desirable -- are answered by the same people and institutions which generated the proposal and the resources in the first place. And their answers are biased. Of course, the dreamer wants to see his dream realized. Of course, the sponsoring institution wants to proceed. And of course, they assume that the project is desirable to the whole of society; it is obviously desirable to them.

---

[1]Christensen, C. Roland, Kenneth Andrews, and Joe Bower. Business Policy: Text and Cases, Fourth Edition. Homewood, IL: Richard Irwin, 1978.

Our current macro-decision model contains no mechanism save serendipity for testing the value judgments of the self-interested institutions which promote the projects. All of us involved in the writing of this volume have based our contributions on our own value systems and our own senses of social responsibility. Whose values should apply to the evaluation of macro-proposals?

Certainly not my own, at least not in isolation. Some of the proposals contained in the second half of this book are absolutely appalling to me. If they were to gain financial support, I should feel constrained to take to the streets, if necessary, to fight them. Better there were some mechanism to resist those projects before they were judged "viable" by some decision-maker, because once that judgment is reached, huge resources are suddenly put to work to convince the public of the project's desirability. By the time the juggernaut begins to move, its mass is enough to crush dissent.

Harris points out that the essence of freedom is the ability to say "no" to challenge. McHale's experience indicates simply saying no is not enough; the people riding the juggernaut are not obligated to listen, and they have a loudspeaker system capable of drowning out the lone negative voice.

Alternatives for a New Macro-Decision System

A few of the writers here have suggested possible mechanisms for wider involvement of all of us in answering the questions about the desirability of proceeding with identified macro-activities. Kozmetsky believes the matter may be safely delegated to established business leaders. Davidson feels the education and training of project managers may hold a key for better macro-decision-making. Harris suggests deliberate allocation of project resources for public education might promote a general feeling of "ownership." Huessy advocates an organized application of social psychology to all macro-proposals, and Kozmetsky the development of a field of social engineering to match the hardware engineering we are now capable of.

It is premature to attempt evaluation of these ideas now. They are young and vulnerable, and deserve the chance to mature before undergoing criticism. Further, they are hardly an exhaustive list of "what we might do" to provide a new, more effective, more efficient, more broadly-based macro-decision system. A possible next step in designing such a system might be turning our attention to development of a comprehensive set of alternative decision-system designs. Given such a set of specific plans, we could subject them to evaluation according to the criteria we have developed here, combine or modify them to remedy weaknesses, and gradually converge on a system design which is both desirable and viable.

In our deliberations here, we have agreed that some form of macro-decision system is highly desirable. Such a system would itself be a macro-activity of tremendous impact on society. Building upon the values of today's populations, it would shape the values of tomorrow's. As those values are applied, they should move us toward new and more constructive applications of large-scale technology. They should point us toward calm consideration of the inherent trade-offs in any macro-proposal -- trade-offs between the present and the future, between labor- and capital-intensity, and between large and small scale.

In our society, bigness is a fact of life. It will not and cannot simply go away because it makes some of us uncomfortable. Bigness is not intrinsically bad; it has as much potential for benevolence as for destruction. How we use it will make the difference, and how we control it will govern how we use it. The challenge facing all of us is the design of a macro-control decision system -- one which considers what "we" want to do and should do, as well as what we might and can do. It must be possible to make the big, beautiful.

Part 2

Cases in Point: Land, Sea and Aerospace

Section A

Transforming the Urban and Rural Environments

Aerial view of the proposed new capital city of Nigeria. The central axis of the city traverses four low, rounded hills, interspersed with minor stream valleys. Important public buildings will be built where the axis crosses the high point of each hill.

Overview

The proposed new capital city of Nigeria has been described as the largest free-standing new town project ever undertaken, with an estimated population of 1.6 million people by the year 2000 and a probable area of 25,000 hectares. Construction is scheduled to begin in 1979, with development to proceed in stages over a 20-year period. The master plan for the new city emphasizes adequate transportation, housing, balanced employment and public service infra-structure, and the use of technologies appropriate to the culture and management capabilities of the country. Many issues have not been resolved, but in view of the commitment of the present government and the degree of popular support, Andrew C. Lemer guesses that the planned capital city of Nigeria, located in the interior in an ethnically neutral area, will indeed become a reality.

Andrew C. Lemer

9. Macro-Project in West Africa: Planning Nigeria's New Capital

After ten years of military rule following the Civil War, Nigeria will soon return to civilian government under a new American-style constitution. The nation's buoyant economy, running on petroleum, is supporting major expansion of new infrastructure and rapid industrial growth. Government planners are facing difficult problems due to accelerating urbanization and extreme poverty.

In this context, Nigeria has decided to embark on construction of a new Federal Capital City (FCC). A consortium of American consulting firms* was selected, through international competition, to plan and initiate design of this major macro-engineering project.

The experience of building a city of this scale is still in the future. We can only discuss the basis and progress of overall master planning which is nearing completion, and make conjectures about lessons to be learned.

Background

Construction of the new Federal capital is scheduled to begin in early 1979, with the principal functions of Federal government to be moved to the FCC by 1986. The city is planned to grow to a population of 1.6 million people by the year 2000,

*International Planning Associates is a consortium of Archisystems, Inc. (a division of the Summa Corp.), Planning Research Corp., and Wallace, McHarg, Roberts and Todd.

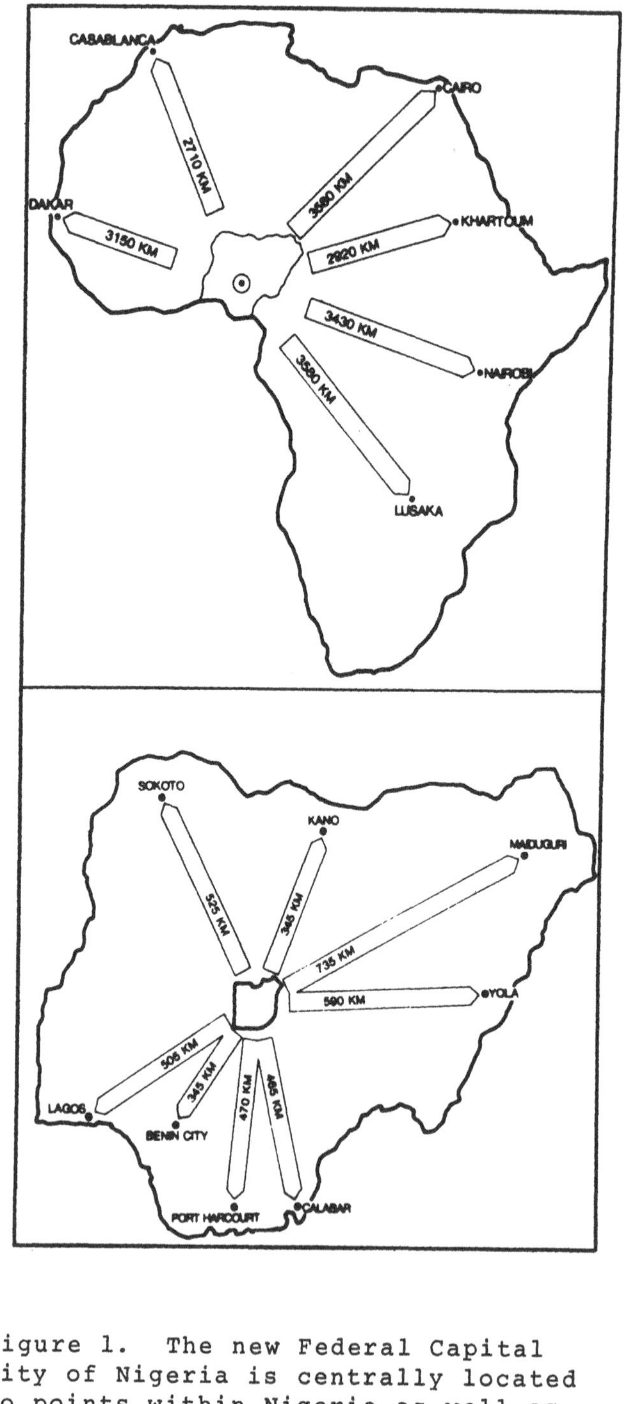

Figure 1. The new Federal Capital City of Nigeria is centrally located to points within Nigeria as well as to the rest of the continent of Africa.

and to occupy an area of approximately 25,000 hectares. The site of the FCC was selected to accomodate an ultimate population of up to 3 million people.

The reasons for moving the capital from Lagos, Africa's largest city, are several: to locate government in a centrally positioned and ethnically neutral area, to escape the chaos of a city which has long-since outgrown the capacity of its infrastructure, and to provide a showcase for national unity and modern African urban development.

At present, the city has no name. While no formal requests have been made or contests run to find a suitable title for Nigeria's new capital, active public interest has been reflected in the proposals for names, which frequently appear in letters to the editors of the nation's principal newspapers.

As with any project of this magnitude, there are serious questions regarding the viability and relevance of the new capital. With the problems of poverty and low quality of life widespread, can the nation afford to concentrate a significant portion of its resources in a single location? At a time when the development of new cities is being questioned in many developed countries, as a reasonable strategy for managing urban growth, can the FCC be expected to meet the ambitious objectives set for it? With the uncertainties of the international petroleum economy and a dynamic domestic political situation, can Nigeria maintain the long-term commitments required to carry the project to a successful completion?

Location of the Capital

The FCC is located within a centrally positioned Federal Capital Territory (FCT). The FCT is under direct administrative control of the Federal Capital Development Authority (FCDA), a specially appointed agency of the Nigerian government, whose chairman holds ministerial rank.

The FCT is situated just north of the humid lowlands of the Niger/Benue Rivers trough, but south of the driest Sahelian portions of Nigera, in an area called the Middle Belt (Figure 1).

The climate is tropical, and exhibits distinct wet and dry seasons. Monthly precipitation is estimated to range from negligible in the November-January period to a peak of 300 mm.

The FCT is generally characterized as a tilted plain rising slowly in elevation from southwest to northeast. Several ranges of low mountains and numerous rocky knobs rise above this plain. These rocky knobs are particularly prominent in the northeastern portions of the FCT, providing a strong visual image and distinctive landmarks for the site of the new capital. These landmarks play a strong role in the proposed FCC plans, providing a setting of appropriate variety, interest, and image for Nigeria's new capital.

The central location of the FCT and the topography of the FCC combine to produce a significant variation in climate as one goes from southwest to northeast parts of the territory. The rainy season is shorter and temperatures somewhat lower in the northeastern areas. Of the several areas of adequate buildable land within the FCT, the proposed site of the FCC offers the best climate.

The FCC Plan -- An Overview

The master plan of the FCC has been developed to be responsive to a wide range of factors:

- The plan to have a single contiguous urban mass is a result of regional land suitability, micro-climatic factors, and a desire to minimize infrastructure costs.

- The decision to have a dominant center rather than multiple centers is based upon a desire to create an appropriate image of importance for the seat of government and to support efficient functioning of government.

- The location of the center was chosen to use prominent natural features of the site as a means of emphasizing the uniqueness of the FCC.

- The curvilinear form of the plan is based upon topography and environmental factors, combined with a desire to maximize transport efficiency.

- The structure of residential sectors and service systems is based upon existing characteristics of Nigerian society, combined with the nation's objectives for its future development.

The most important single factor in determining the physical organization of the FCC is transportation structure. Two parallel transit spines are brought together at the central core, which is the main government, institutional, and commercial complex. Bounded by peripheral highways, the FCC follows these corridors, assuming a form which is essentially a "wedges and corridors" radial plan deformed to a linear shape. The great majority of residents will be within walking distance of the transit system.

Along the two corridors, residential areas are organized into distinct sectors. Each sector, with a population of between 75,000 and 200,000 people, is intended to be a mini-city providing its residents with local employment opportunities, community and commercial services. Each sector is further divided into districts, their sizes determined by population necessary to support schools, health care facilities, markets, etc.

Separating the corridors and surrounding much of the city are reserved urban parks and open space. Stream valleys comprise an important portion of this reserved open space, providing natural drainage, and aesthetic benefits.

Industrial development, which we estimate to be a relatively small element in this government/services-based city, is concentrated in estates located at lower elevations than the residential and commercial areas. Intersections of the road network provide these areas with good accessibility. An international airport is located further to the west and linked to the city center by a limited access expressway.

Housing

As one of the most basic of human needs and

the most direct point of contact between the FCC and its residents, housing is a major concern in the FCC. The housing program attempts to strike a balance in the conflict between the standards which the government wishes to provide its citizens and the costs which the FCC's residents and the government can afford.

As with the use of modular sectors to provide flexibility in the overall plan, the housing program is intended to respond to a variety of alternative institutional and financial approaches to implementation. Land is reserved in the plan to permit expansion of the numbers of housing units and sizes of individual units -- as incomes rise -- without having resultant uncontrolled sprawl of the FCC.

Within the individual housing units, it is assumed that finished work will be reduced and self-help -- "sweat equity"-- will be employed for upgrading. However, a detailed housing design has not yet been undertaken, and it is likely that the FCC's housing program will undergo review and revision as development proceeds.

Public Services

After basic shelter and adequate and balanced employment, provision of access to good quality public services is a major aim of planning. While actual quality of service delivery is dependent upon the organizations having final responsibility for operations, service facilities have been sized and located for accessibility.

Schools have been used as foci for residential organization and community activities. Implementation of universal primary education decrees and assumed increases in enrollment rates at all levels make education a major land use within the FCC.

Health services are to be provided in an integrated system, from a major hospital complex with teaching facilities down through specialty clinics to frequent local clinics and baby care facilities. Libraries, post and telegraph facilities, police and fire services, and cinemas, theaters, and other local cultural and recreational facilities are distributed in accordance with Nigerian practice and recommended international standards.

As the center of the nation, the FCC will include a number of unique cultural, recreational, and governmental facilities, many of monumental significance. It is expected that these facilities, including a national stadium, library, capital and executive residence, will be designed by Nigerian architects. Consideration is being given to holding an invitational competition for architectural design for the central area of the city.

Implementation and Incremental Growth

Development of the FCC is planned to occur in stages over a 20-year period. Staging is based upon the residential sectors, allowing flexibility to speed or slow the process without upsetting the functional integrity and viability of the FCC as a whole. Population growth, and thus plans for implementation, are based upon projections of Federal government employment and policy toward decentralization.

Imbedded in the population projection of 1.6 million people by the year 2000 and related to a city size estimate of 25,000 hectares are a large number of assumptions regarding labor force participation, family size, relative incomes, and employment characteristics. An important factor in the FCC's growth will be the "informal" sector of its economy, the artisans, vendors, domestic servants, part-time workers, etc., who comprise a major portion of the work-force in many developing nations.

A close partnership of public and private sectors of the Nigerian economy will be required in the FCC's development. While the new city will be built to serve the functions of government, a large proportion of the FCC's costs will be associated with activities of non-governmental business and private individuals.

The plan reflects a principle that the proper role of government in developing the FCC is to provide necessary infrastructure, social overhead, and economic base to support natural growth and development. Much of the investment which will occur in the course of this natural growth will be made by the private sector.

Project Management

In general, a "fast track" approach is required for the FCC in planning and design as well as in construction. Construction will begin on the initial residential district while many of the large-scale planning issues are still being resolved.

The Federal Capital Development Authority (FCDA) was created to bring the concept of a new capital to actuality. This ministerial level body has full governmental authority to develop and administer the FCC, and is limited only by its own technical and managerial capabilities, and by yet to be established Federal budget constraints. The FCDA is currently undergoing rapid increases of personnel and is faced with difficult decisions regarding basic approach to continuing project development and construction management.

Logistics

While the city for 1.6 million people will be a major challenge to complete, we feel that the most difficult goal is that of moving government by 1986. It is planned to have a population of 150,000 people by that time, housed in permanent facilities. The initial infrastructure and housing required will represent a heavy demand on the construction sector. Peak demands for materials, equipment, labor, and transport services are projected to occur during the 1982-84 period.

Through 1981, we estimate that the existing system for supply of construction materials will be adequate to avoid major shortages and inflationary pressures. However, most of these materials are imported, and as FCC material demands increase, we feel that some mechanism for advanced procurement and stockpiling will be necessary.

Similarly, the initial workforce is estimated to be available, and might be housed in temporary quarters in nearby villages. However, as the labor force grows to its estimated peak of 44,000, housing will be required on-site. We anticipate that initial construction will be permanent housing units, which may be used for this purpose.

While it is estimated that the existing transport system has adequate capacity to meet demands during this period, dedicated port facilities and coordination of trucking are essential if the delays endemic to Lagos are to be avoided. Over the longer term, barging of materials up the Niger River appears to be a most promising means of supplying the FCC.

Is a New City a Good Idea?

It is estimated that the urban population of tropical Africa is growing at a rate in excess of 5 percent per year, significantly faster than the overall population growth rate of 3 percent per year. The impact of the coming "astonishing population increase" will be felt mainly in existing cities.[1]

Lagos, the largest city in black Africa, is an early and graphic case in point. The population of Greater Lagos, approaching 3.5 million people, will have increased by more than the 75 percent projected for the 1970-1980 period. The nation as a whole will have grown approximately 32 percent.

This growth has not been accompanied by comparable expansion of urban infrastructure and services. Lagos is "well on its way to becoming the Calcutta of Africa."[2]

Yet, in spite of such massive urban growth and expectations for its continuation, the population of tropical Africa is still overwhelmingly rural. It is estimated that only 11 percent of the approximately 242 million people in the 35 countries of the region lived in settlements of 20,000 or more individuals.

---

[1]International Planning Association. Draft Master Plan for the Federal Capital. Washington and Lagos, September 1978.

[2]Rosser, C. Urbanization in Tropical Africa: A Demographic Introduction, International Urbanization Survey, New York, Ford Foundation, 1972.

Appeal of a New City

Hence, many of the people coming to Lagos and the other major cities of the region are likely to be experiencing their first contact with any form of urban life. Lagos, with its sprawling slums, erratic water and electricity is a clearly ineffective mechanism for productive integration of these migrants into the urban economy.

The concept of a new city is thus especially appealing. The FCC is a potential magnet to attract migration from Lagos, while providing relatively immediate relief for government operations severely hampered by the inefficiencies of Lagos' congestion.

But even if one accepts the desirability of a new city, the size is still a major issue. In spite of extensive research, no clear-cut relation between urban sectoral efficiency and city size has been proven. It has, however, been argued that larger cities are required to support large services sectors, particularly the high-order "informal" sector typical of many LDC cities.[3]

It is estimated that in the year 2000 more than 40 percent of the total employment of FCC will be in this informal sector. Overall labor force participation is estimated at 52 percent of the population.

If the experience of Lagos -- and to a somewhat lesser extent, other cities in Nigeria -- is applicable, it is unavoidable that the FCC will grow large. Indeed, it is hoped that the FCC will shift migration and urbanization more toward the central areas of the nation, into the "Middle Belt" where development has lagged behind both north and south.

The challenge will be to control the early years of the FCC's growth. Migration to the job opportunities of future construction will have to be restricted to assure that population does not expand beyond the capacity of infrastructure.

---

[3]Richardson, H.W. City Size and National Spatial Strategies in Developing Countries, Staff Working Paper 252. Washington: World Bank, 1977.

Balanced Planning and Appropriate Technology

A broad view must be taken of what infrastructure is required to support population growth. The public services of the FCC are as important as the physical systems of transport, water supply, and sewerage. Balanced development of these social systems represents an early investment in human resources. This investment will be extended to education. A program of manpower training is planned to provide construction labor for the city's development.

Throughout planning of both physical and social systems of the FCC, "appropriate technology" has been a matter of continuing concern. While a new city provides an outstanding opportunity to apply new ideas and emerging technologies, lack of technical sophistication in the society as a whole dramatically increases the risks of total failure of these new ways. This lack of sophistication is manifested in the absence of personnel trained to operate and maintain advanced systems and inexperience of the population in the use of such systems.

If the FCC can demonstrate the value and applicability of new technologies for the West African city, it will have served a valuable function, one that is unlikely to be possible with existing cities. However, a key to success will be judicious blending of new technology with old traditions. For example, the transit spines which are the major structural feature of the FCC are planned to be busways -- paved roadways with use restricted to public transit vehicles. Such restricted-use roadways are new to Nigeria, and it is not unlikely that stringent measures will be required initially to assure that access to the roadway is indeed limited.

For example, modern expressways in Lagos are often obstructed by cattle crossing the roadway. Barriers have been constructed in the median of some roads in an unsuccessful attempt to encourage people to cross at available pedestrian bridges or underpasses rather than crossing active traffic lanes.

The vehicles using the busways will not necessarily be only large transit buses. "Mammy wagons"

Figure 2. Artist's concept of the FCC's parliamentary buildings from the Mall.

and other traditional forms of transit, which comprise an extremely important part of the urban transport system in Nigeria, should play a major role in the FCC.

In contrast to the transportation system, advanced technology is planned as a means of minimizing maintenance and operating difficulties in telecommunications. It is anticipated that modular digital solid-state systems and fiber optic transmission may be employed. Centralization of maintenance and network management will then permit a small group of trained technicians to care for the FCC telecommunications.

Symbol for the Nation

As a potential symbol for Nigeria's future, the new capital follows a long tradition. Canberra, Brasilia, and Washington are national capitals which have grown from grand plans placed on virgin land.

As Eldredge states in his study of world capitals,[4] the capital "bears the symbolic torch for the entire nation -- the image the chosen city creates is a powerful factor in national status and internal self-conception." The FCC is sited and planned to play this role.

Image is initially provided by the landscape. The FCC's central area will be aligned to focus on Aso Hill. Major symbolic architecture--monuments, legislature, etc.--are to be located on low hills along the axis (Figure 2).

Image will, of course, be provided primarily by design of individual buildings and sub-areas within the FCC. It is envisioned that there will be an international competition for design of the central area.

In addition to the visual image, the FCC will be symbolic in the standards of urban life it is intended to provide. Planning decisions have been

---

[4]Eldredge, H.W., ed. World Capitals: Toward Guided Urbanization. Garden City, NJ: Anchor Press/ Doubleday, 1975.

made to try to maintain a balance between quality and affordability, in an effort to assure that the plan will be implementable. A new city avoids the physical-psychological obstacles to implementation which an existing urban structure may present.

The FCC as Feasible Macro-Project

Nigeria's Federal Capital will be the largest free-standing new town project ever undertaken. However, the problems which a project of this scale might be expected to experience are in large measure mitigated by the time period allotted to its implementation.

As discussed previously, the most difficult years will be the period between now and 1986. To produce a functioning urban environment to house the nation's capital and 150,000 residents requires construction of quantities of infrastructure, buildings, and monuments out of proportion to the population as a part of the total 1.6 million people scheduled for the year 2000.

Only preliminary estimates of costs and financing requirements have so far been made, and if the lessons of other major projects are relevant, these initial estimates are likely to be low. Yet even allowing for cost growth, the Federal government's share of annual investment in the FCC appears to be well within the limits of the nation's current budgets and practical capacity. Any future real increases in oil revenues would assure that development of the FCC can be completed with no strain on the economy.

There may be problems in the early years with supply of labor and materials. Congestion of the transport system and the desire to minimize importation of construction materials and skilled labor could slow initial progress. As discussed previously, logistics planning has been undertaken, and with careful management, these problems can be minimized.

Indeed, it is development of adequate management capacity which may be the principal challenge to successful development of the FCC. Estimates of personnel requirements indicate that substantial growth of FCDA procurement and contract administration staff will be required if project management is to be provided "in-house." There has already been a major increase of architectural and engineering personnel, recruited primarily from several eastern European and Asian countries.

As with the problems of logistics, there are a number of ways in which these potential management difficulties may be solved. The apparent commitment of the government to the FCC suggests that such problems will, in fact, be solved. Popular interest, as reflected in newspaper editorials and experience of members of the planning team in Nigeria, appears to indicate that this commitment will survive the transition to civilian rule. It thus seems quite possible that the plan for a new Federal Capital of Nigeria will indeed become a reality.

Overview

One possible solution to the problems of the increase in scale of contemporary society would be to utilize the surface of the sea for living space and to make greater use of our waterways for marine mass transit. According to John P. Craven, the cultural advantages of water-based cities have been demonstrated in the neolithic lake societies, and the technological feasibility of ocean cities has been demonstrated by the numerous varieties of offshore platforms produced in recent years. The increased use of marine mass transit on rivers and across oceans would afford designers an opportunity to furnish examples of ideal urban characteristics. Now that technical feasibility has been demonstrated, it only remains for a favorable social-political climate to emerge for ocean cities and urban marine mass transit to be granted a fair trial.

John P. Craven

10. Cities of the Future: The Maritime Dimension

An overriding and apparently inexorable feature of the history of societal organization is the increase in scale. The size of cities, of buildings, of athletic stadiums, of theatres, of thoroughfares, of homes, of shopping centers; the volume which society allocates to each individual at home, at work, in the office; the size and velocity of vehicles for transport of passengers, the number of passengers per vehicle; the volume of communication, the rate of transmittal and the distances over which they are transmitted and the number of recipients of each communication; all show monotonic increases. Only a few isolated exceptions to this trend can be cited such as the Caracalla baths of Rome, the 2,000-passenger ferry boats, or the estates of the affluent. This growth in scale is so apparent and in recent times so accelerated that analysts have succumbed to the temptation to represent this growth by empirical equations of the form Ce^{at} and to thereby extrapolate a growth which would constitute a clear and present danger.

In the face of this perception or misperception, compelling arguments have been made that "small is beautiful." Indeed, a number of small groups of individuals who, failing to perceive that they are, in fact, numbered among the world's most affluent, have been able to afford the imitation of the lifestyle of a previous generation when the planet Earth had a much smaller population in relation to its total resources.

If we recognize that e is a constant of mathematical convenience ($\int_0^1 e^x dx = e^x$) and not a constant of nature, that growth processes are in general non-linear, and that growth rates are very different for different functions of living, then we are ready to disregard dangers which are the product of over-simplification and ask more exciting and relevant questions. We should note that many of life's functions are not amenable to changes in scale and are not likely to change for many decades or even centuries. These include but are not limited to life expectancy, athletic capability, the rate of speaking, the rate of writing, the capacity of the brain to receive information in rate or total quantity, the period of gestation, the years of childhood and adolescence, the productive years, the average size of the family, of social groups, clubs and fraternal orders, etc. Although most of these factors have been extended or expanded, they represent nearly constant and size-limiting factors in the scale of society.

This paper, therefore, examines the more exciting question: How big can the other scales of society be in order to benefit from the economies of scale and still be beautiful? The word "beautiful" is used in the sense of aesthetically and sociologically satisfying to the populations of such societies.

We shall argue in this paper that at least one order of major increase in the scale of society (the size of cities, the population of earth, the size of vehicles of transport, etc.) can be accomplished while restoring or establishing a "beautiful" society. The argument depends for its validity on the utilization of the surface of the sea, the bays, the rivers, and tributaries of the world as usable space. The argument is equally dependent on the assertion that technological breakthroughs have occurred which will eliminate the two major inhibitors to living on the sea surface. These are the perils of the sea associated with storm and aggravated sea state, and motion sickness and psychological discomfort associated with sensible motion of the world in which we live.

Before we can make this argument, we must define what we mean by a "beautiful" world.

Although it is doubtful that agreement can or will be obtained for any such definition, an attempt has been made by this writer and the Japanese architect Kiyonori Kikutake to define the design requirements for an urban society and, more specifically, the city.

In broadest terms, an ideal city will: 1. meet minimum requirements for the mechanics of living, 2. provide protection against natural and man-made depredations or aversive stimuli, and 3. provide ample opportunities for each individual to satisfy his or her needs for reward, i.e., the provision of abundant opportunities for the widest range of positive stimuli. If we examine each of these factors in some detail, we may see those which are amenable to satisfaction by increases in the scale of society, and those which are not.

The minimum requirements for survival are well known and all city planners and designers are careful to include these factors in their layout and plan. They include: an economic supply of water, food, clothing, and shelter for all the inhabitants; non-polluting, rapid, comfortable, flexible, and immediate transportation for people who may be carrying from 40 to 80 pounds of goods; non-polluting and non-obstructive transportation of goods in quantities in excess of 80 pounds; telephone, radio, television, printed media, mail, accounting, and computational communications; non-polluting, efficient, and effortless disposal of wastes--human, solid, liquid, organic, toxic, and non-toxic.

These minimum requirements apply equally well to a prison community, an army, a ship's crew, space voyagers, or any other aggregation or colony of individuals who are destined to live out a normal span of life. To the extent that a city is a voluntary association of individuals, satisfaction of life-support criteria is not sufficient to insure its viability. The Thorndyke-Skinner model of human behavior tells us that people will repeat behavior and performance when that behavior (even though accidentally or randomly initiated) is rewarded. This model also tells us that although people will adapt to, ignore, or even find reward in mild irritants, they will react strongly to destructively punitive phenomena or aversive stimuli. The avoidance of aversive stimuli is thus another mandatory criterion for city design. These avoidances include: freedom from pollution--atmospheric, water, land, noise, and time; protection from fire, flood, tsunami, earthquake, hurricane, tornado, volcano, and other natural disasters; military defense and protection against crime, civil disorder, and civil disaster; protection against sickness, disease, epidemics, and bodily injury; provision of adequate public and private health and hospital facilities.

Fulfillment of these design criteria will make the city livable but not lovable. It is the abundance and availability of positive reinforcements that give a city charm and desirability, that make it attractive and induce people to live in its environs as a matter of choice. The list of potential positive reinforcements is long and has too often been overlooked by city designers. Our (Kikutake, Craven) list is as follows:

- Opportunities to derive adequate economic income

- Opportunities for social affiliation in recreation, organization, sports, spectator sports, conversation, theatre, arts, music, etc.

- Opportunities for achievement in organizations, in business, intellectual, and physical activities, in culture, the arts, etc.

- Opportunities to engage in dominance and/or submission with family or selected associates, or with pets or inanimate hobby objects

- Opportunities to engage in dogma and independent belief with family or selected associates, or with pets and hobbies

- Opportunities for privacy, individually, in pairs, or in small groups

- Opportunities for afferent reinforcement through experiencing spacious and unscarred natural landscapes and environments

- Opportunities for experiencing variety and the freedom to select or alter one's local environment

- Preservation of similar opportunity for posterity

Given this large number of conditions, one may well ask whether there exists any reasonable configuration which will permit all of the reward stimuli to flourish and at the same time protect against all of the aversive stimuli and meet minimum requirements for the mechanics of living. Quite obviously many of the reward goals are denied by modern urban-suburban living.

For example, the low density in the suburbs has made social affiliation on a non-planned, casual basis extremely difficult. In the city proper, crime in the streets is a real and unsolved problem, not only for the victims of this aversive stimulus but as an expression of unfulfilled needs on the part of the individuals engaged in this anti-social behavior. The fear of crime makes prisoners out of many apartment-dwellers, particularly the aged and the female, destroying their own opportunities for socialization. The need for companionship and domination is often expressed in the ownership of pets which, in turn, creates aesthetic and public health problems in the city streets. The security-conscious nature of the apartment complex, which denies door-to-door solicitation and is restrictive of unsolicited mail communication, isolates the individual from dogma-resolving organizations such as the church and various forms of the political process. Many other examples can be cited, but the reader who has any experience within the city can cite countless situations and times when these idealized reward structures are unavailable or are defeated.

Perhaps these reward goals are unachievable in the city and thus unrealistic or inappropriate for design purposes. If we believe that the city is organic (i.e., that its formation and growth processes are basically unplanned), then there will have been an attempt or attempts to adapt to an optimum form of social existence. If we can find (and we can) examples of stable, satisfied communities whose existence has not been transient, then we have an empirical test of these design principles. Certainly, the evolutionary process which produced these communities would have satisfied the suggested principles of city design if the principles are indeed appropriate and correct.

If we seek a solution at the smallest scale of society, we can find it in man's Neolithic lake societies. As proof of the fundamental and organic nature of these societies, we note that Neolithic communities of nearly identical form and size grew and flourished around the world in temperate and semi-tropical zones. We can understand the pattern of development as man the hunter followed the small game to the lake shore. There the seeds of the fruits consumed by the herbivores were deposited in their feces on the fertile muds of the edge of the lake to initiate an annual cycle of seed time and harvest. Nature thus provided the first agrarian hold on roving man, and communities developed on the shoreline. But shoreline communities did not meet the needs of society, (at least not as outlined in this paper) and in order to survive, the community had to move to the center of the lake. The great historian Herodotus describes one of these communities as follows:

> Their manner of living is the following. Platforms supported upon tall piles stand in the middle of the lake, which are approached from the land by a single narrow bridge. At the first, the piles which bear up the platforms were fixed in their places by the whole body of the citizens, but since that time the custom which has prevailed about fixing them is this: they are brought from a hill called Orbelus, and every man drives in three for each wife that he marries. Now the men have all many wives apiece, and this is the way in which they live. Each has his own hut, wherein he dwells, upon one of the platforms and each has also a trap door giving access to the lake beneath; and their wont is to tie their baby children by the foot with a string, to save them from rolling into the water. They feed their horses (sic)* and their other beasts upon fish, which abound in the lake to

*Probably "pigs", rather than "horses".

such a degree, that a man has only to open his trap door and to let down a basket by a rope into the water, and then to wait a very short time, when up he draws it quite full of them. The fish are of two kinds, which they call the paprax and the tilon.

The description would be equally valid for any one of the Neolithic lake societies whether in Britain, or Switzerland, or Etruria, or Africa, or Peru, or Thailand, or the Philippines. Examination of these early communities reveals how closely they meet the design principles of our utopian city. The economic supply of food was obtained from the game of the high hills and the abundant fish supply of the lake; water for drinking came from mountain streams continuously flowing from the winter crop of snow; clothing from animal hides; and shelter from the logs of tall timbers felled by acts of nature. Land transport was also facilitated by topography, it being necessary only to drag the large game downhill to the water's edge, there to transport by coracle or raft to the pile dwellings. Communication by drum or smoke signal was adequate for the day, and disposal of waste was simply accomplished by squatting over the platform opening and dropping into the lake. These simple cultures knew, as modern societies know, that drinking the waters of the lake was tabu; but they also knew what the environmental superstitions of modern man have suppressed: that swimming in waters enriched by biological waste presents no hazard to the health. These same untreated wastes were rapidly converted to fish more than fit for human consumption. Thus, the Neolithic lake community met the minimum requirements for survival in a handsome manner.

Not all of the aversive stimuli of nature and man were avoided, but these communities were remarkably free from many that afflict modern man. We have already seen how one or two simple tabus freed them of most of the pollution problems which so afflict our day. Less obvious is the protection from fire, flood, and earthquake afforded by the waters of the lake from their properties as an extinguisher, as a flood control reservoir, and as a decoupling medium for earth tremor. The configuration of the community was ideal for defense, particularly against invaders having no knowledge of the art of swimming or seamanship. One can see the attacker now, launching clumsy and ineffective log rafts which are quickly tipped by the defenders who adroitly grapple with the invader in the waters of the lake. How quickly the attacker's grip becomes that of a drowning victim rather than that of a warrior in battle. One can see the impossibility of a blockade, since by stealth of night the entire periphery of the lake is available as a location for landing and embarkation. Thus, defense against the invader was simple. Crime and civil disorder were also of minor social import as the result of the rigid tribal and family structures of these simple communities which visited punishment on the iniquitous rapidly, simply,and rough-handedly.

What really made these early-day Walden ponds a stable and happy culture was the existence of many opportunities for reward. The organization of the platforms provided an ideal configuration for social affiliation. Dwellings were in general located at the periphery, and community and family gatherings were in the center of the platform configuration. Thus, the assembly process from the low density family unit to the high-density community meeting was quickly and easily accomplished. At the same time, dwellers on the periphery could look outward and lift their eyes unto the hills, experiencing the wonders of spacious and unscarred natural landscapes, the snows of winter, the greenery of summer, the foliage of fall. Opportunities for achievement in the hunt, or in sports in clearings on the shore, or within the tribal structure were abundant, and the community was bound spiritually by religious dogma and symbolism.

When these communities failed, it was due to their inability to preserve the opportunities of the societies for their posterity. As Herodotus pointed out, each man had many wives. Thus, each man had many children, and with the growth of population came the increase of platforms until the lake was covered, and the vital functions of defense, transportation, fishing, aesthetic reinforcement, etc., were severely compromised. The solution for many of these societies was exile over the rim of the crater which had given birth to the lake, onto the streams that flowed inexorably down to the sea. Thus, pieces of civilization were forced from their stable and tranquil land-sea configuration to start life anew in the coastal zone.

Before we leave these tranquil societies (many of which survive to the present day) we note that these communities can be represented symbolically by a three-dimensional configuration having the following basic characteristics: the land is reserved for agricultural, recreational, and low population density functions of the society; the surface of the streams and the large mass of water is reserved for transportation. High density living takes place in three-dimensional beehive-type structures located above the water surface with lowest density on the outer periphery and highest density in the center. The technological sophistication of the society is to be found in these structures. The waters below the surface are utilized for industrial functions, for cooling, for processing of wastes, for manufacture of protein. The total configuration provides simultaneously the natural configuration for defense and a configuration which, blending with the environment, requires little in the way of maintenance and upkeep.

We may now ask whether modern technology can replicate the Neolithic model at a larger scale and, if so, what may be the current economically feasible limit. We know that our society enjoys the aggregation of 100,000 or more people in a single stadium for the reinforcement that mass participation in spectator sports provides. Thus, concentrations as high as 5,000 people per acre are desired. On the other hand, for open space enjoyment, concentrations of less than one person per acre are desired.

These density dilemmas can be resolved by the physical organization of society in a three-dimensional community which is purely pedestrian or which at most employs simple people-movers (escalators, moving walks, golf carts, etc.). If, for example, we desire to collect an audience of, say, 10,000 people for a concert from a population of, say, 100,000 who are distributed in a circumferential belt having the conference hall as a center, then one must travel some four miles before the suburban density of 10 to 20 people per acre is obtained. On the other hand, if a truly three-dimensional city having a height of 20 stories were to be constructed, then this density would be achieved at the healthy walking distance of one mile.

To understand this concept more fully, we should note those activities which require a view, those which do not, those which admit of high density, and those which do not. In the first category are theaters, concert halls, basketball arenas, even baseball and football stadiums, churches, classrooms, political rally arenas, etc. Of lesser density, but nevertheless high, are department stores, grocery stores, cleaners, service establishments, restaurants, hospitals, medical centers, business offices, barber shops, beauty shops, etc. Lower density goes with living spaces, and lowest density with parks and airports.

Conceptually, this leads to a beehive concept in which living spaces are on the outer skin, nonvisible one to the other, and each with a view of an unencumbered environment over a solid angle porjected from the viewer and limited only by the collimation required for privacy. The individual travels internally into the hive toward successively denser areas of population: at the outermost level for socialization in plazas and courtyards; at an inner level for shopping and services; at a further inner level for offices and business and vocation; and at the core for entertainment and convocation.

To meet low-density needs of the population, the resident of the complex travels externally from the beehive to parks and open spaces to a transportation system which can carry him to the distant parts of the environmental park which surrounds his high-density complex. Ideally, this transportation system should be placed below the "beehive" so that it is equally accessible to all citizens; or (as is done at Disneyland) a peripheral transportation system could be provided to carry individuals to the desired outbound radial.

The model as thus far described leaves out the space and locational requirements for industry, waste disposal, communication, power, warehousing, movement of freight. If these could be located below the beehive, then an arrangement idealization is obtained which in theory would allow the full design goals of the city to be met. This concept has been explored by Paolo Soleri in his arcology concept.

The realization of the Soleri concept presupposes that the arcologies which he has designed are structurally feasible within the current state of the art and that a transportation system is provided which can move people from the high density core of the city to the low density parkland periphery. The realization further presupposes that the energy and capital costs will be low enough so that the community is economically viable in competition with other forms of social organization.

If we first examine the transportation, we note that technology now allows three major alternatives: the motor car, the fixed guideway rail system, and waterborne riverine and canal transport. Despite its apparent flexibility, we note that the automobile fails to permit the high density aggregation which is desired for spectator events and it equally denies the opportunities for the low population densities required for wilderness enjoyment. This paradox results from the fundamental character of automotive transport. If the automobile is employed (as it is) for transport to high population density complexes such as athletic stadiums, the attempt to achieve high density is defeated by the extensive fields of parking lots and networks of highway access systems which are required to accommodate the population to be assembled. This limits population density to not more than 500 people per acre. It further requires a separation between shopping centers, theatres, and stadiums so that each may accommodate its own burden of automobiles during the period of maximum utilization. As a consequence, the time pollution resulting from transportation to each complex limits the number of activities which may be enjoyed in any given interval of time.

At the other end of the scale, access to each wilderness area requires the installation of highways, power lines, communication lines, etc. -- a total umbilical package whose economic cost cannot be justified by the low density use. Soon the installed umbilical is loaded with parasitic attachments along its entire length until the maximum economic utility of the umbilical is realized. At this point, the desired low density of population in the wilderness is lost (we might call this the Yellowstone Park phenomenon).

The replacement of the automobile by fixed guideway mass transport systems will be effective in obtaining higher population densities. The effectiveness of the New York City subway system in this regard is undoubted. For example, with the combined resources of the BMT, IRT, 6th Avenue, and 8th Avenue subway systems, it is perfectly feasible to move 100,000 or more passengers per hour into and out of the Times Square area of New York. This would be accomplished without obstruction of the street level surface by any vehicle. On the other hand, the higher lineal cost of fixed guideways makes the use of this form of transport into wilderness areas one of extremely high cost. Nonetheless we must conclude that the use of fixed guideway subway transit systems will satisfy the arcology concept at a scale beyond that now realized in the highest density cities.

If we examine the third alternative of utilization of existing water courses and canals supplemented with a modicum of canal construction, we are surprised to find that they are plentiful, occur in the very center of the areas of highest population, and are connected by means of bays, estuarine, or coastline transit to land masses having little or no population density. This "free" and interconnected fixed guideway system has the further advantage that vehicles of high passenger capacity are easily borne on the water surface. As a consequence, the limitations of the automotive and the fixed guideway system are not present. If, for example, it is desired that transport to wilderness areas be effectuated, the total capital investment is simply that for the most rudimentary of landings.

In the high density portions of our present city complexes, the existence of waterways is merely a concomitant of the drainage which is required for the protection of these high density areas.

To understand how convenient and effective these natural corridors can be, the population density which is available to them in the American city should be quantified. The author has analyzed a number of cities in terms of their ability to benefit from marine mass transit. Population and other demographic data were taken from the Atlas of Twenty American Cities. The usable waterways were identified, and one-quarter and one-half mile swaths were drawn on either side of the waterway. The population within these swaths was then identified and calculated. Substantial percentages of the total population were found to live within these corridors, as follows:

Table 1

	Percentage of Population Living on Either Side of Usable Waterways	
	1/4 Mile Swath	1/2 Mile Swath
New Orleans	27	49
Seattle	25	45
New York	19	39
Honolulu	22	39
Boston	18	34
Philadelphia	12	28
Chicago	9	21
San Francisco	9	18
Cleveland	4	12
Washington, DC	4	12

The population's access to these waterways compares favorably with--and is in many cases superior to--the access to carefully designed subway systems. Of particular significance is that the population density per lineal mile is high and, in most cases, quite uniform along the waterway. A comparison of New York and Boston systems shows a population density of 79,000 per mile* for New York subway systems; 25,000 per mile for New York City waterways; 19,500 per mile for Boston waterways; and 13,600 per mile for Boston subways. (Figure 1 shows accessibility to New York waterways.)

*This figure (79,000) is too high in terms of "population per mile of subway systems" because for about 70 percent of the distance, more than one subway cuts through the swath.

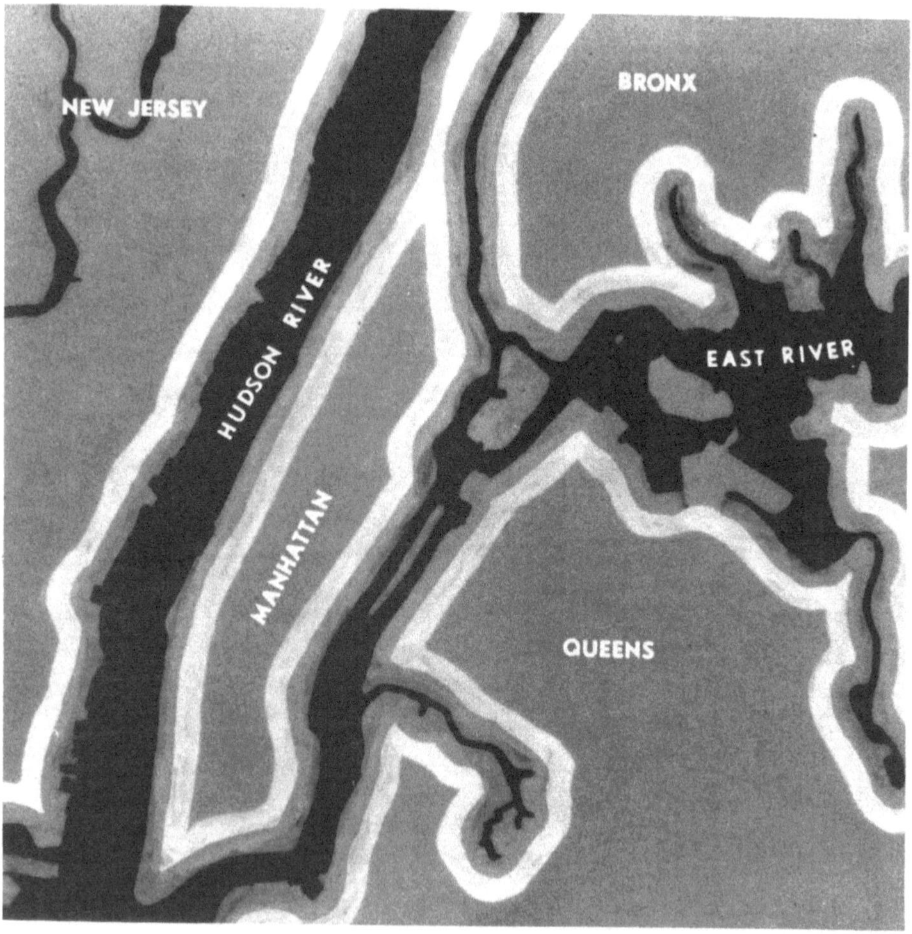

Figure 1. Accessibility of New York to waterways usable for marine mass transit (¼- and ½-mile swaths).

The evidence is quite conclusive that existing waterways have the potential to serve as conduits for public transportation in vital transportation corridors. The utilization of the waterways for revitalization of the entrepot functions of the city appears attractive for many cities. Integration of container barges into the highway networks is relatively easy because of the ubiquitous nature of the net. Revival of the barge for railroad car transport is feasible (should there be any railroads financially able to construct them). Even a cursory examination of subway and fixed guideway systems indicates that nearly all have stations on or near the waterfront in locations that would make interchange construction relatively economical. Interchange with airports is relatively easy in many cities. As previously indicated, and for a multiplicity of reasons which include noise pollution abatement, glide path considerations, availability of filled land, etc., many city airports are located at the edge of the ocean or alongside a major waterway. Thus airports in New York, Philadelphia, Washington, San Francisco, Cleveland, Los Angeles, Boston, Honolulu, have immediate access to the water network.

Given the convenient location of most waterways in the modern city, the next question is, of course, the availability of modern water-based vehicles to act as the couriers on these systems. A fairly substantial number of such vehicles have been developed. Unfortunately, many are unavailable in the United States because of the prohibitions of the Jones Act and the absence of an adequate small-craft shipbuilding capability. It is for this reason that many in the United States are unaware of the revolution in small-craft design and development that has taken place in the past two decades.

Given conveniently located waterways, the next question concerns the availability of modern water-based vehicles which can carry people in quantities to truly qualify as "mass transit," and which have a ride quality in the heaviest of seas in which they will operate such that motion sickness does not result. There are such vehicles. Unfortunately, the regulatory and socio-economic processes of the United States have exiled all of these craft from our shores. The major break-

Figure 2. Mitsui Engineering & Shipbuilding Co., Ltd. Semi-Submersible Passenger Ferry.

through in marine ferries is the semi-submersible platform ferry. This craft, initially developed and currently ignored by the United States Navy, is being introduced into commercial passenger ferry service by Mitsui Engineering and Shipbuilding Co., Ltd. (Figure 2).

The craft consists of twin submarine hulls whose draft is adjustable by ballasting. Twin struts pierce the surface and support a box-like platform above the free surface. The craft has a number of significant characteristics. The chief one, of course, is stability at any speed in a seaway such that motion sickness will rarely occur, and on those rare occasions, the sickness will probably be more psychological than physical. Because the hulls are well below the free surface, the wave drag on the craft is reduced, and a significant speed advantage for a given power can be obtained. Another major advantage is the ease with which it can change draft to accommodate shallow water or to mate with a pier. As such, the craft can play a major role as an interface between canal and river systems and systems which must operate in open and exposed waters. A major advantage may be that of all river craft, the semi-submersible is perhaps the most suitable for roll-on roll-off movement of automobiles. Because of the shape of the hull and because of the inability to substantially change draft, all of the other vehicles are inappropriate for this function. The hull shape makes location and stowage of the vehicles for transit a cumbersome and slow process, and the inability to change draft requires a complex and cumbersome loading ramp facility. The upper structure of the semi-submersible can be arbitrary in shape and is in general box-like or shaped like a parking lot or a multi-tiered parking lot. Regardless of the load, adjustment of ship's draft through movement of ballast will permit the semi-submersible to mate with the access road.

The semi-submersible can be built in any size. Thus, 2,000-passenger craft having displacements of less than one ton per passenger are feasible for conventional ferry operation anywhere on the world's oceans. At modest speeds, the energy cost per passenger per mile will be among the lowest of transportation costs for any mode of transport. Low in cost of production, low in use of energy, high in passenger load capability, and competitive with land mass transit in speed, the semi-submersible represents a fulfillment of the economies of scale and an opportunity to expand by an order of magnitude the passenger-moving capabilities of mass transit systems.

A complement to the semi-submersible is the high-speed hydrofoil. The most sophisticated embodiment of this form of craft is the Boeing hydrofoil. This craft literally flies in the water on underwater foils. The angle of attack of the foils is adjusted to compensate for the waves and a comfortable ride is attained in all but the highest sea states. The 300-ton, 200-300 passenger craft was technically successful but financially unsuccessful in inter-island service in Hawaii. Overseas, they are financially and technically successful, particularly on the run between Hong Kong and Macao. The dollar deterioration has made them attractive for use in Japan.

On sheltered waters many alternatives are available. The long-time standard of comfort and convenience and volume transport remains the New York City ferry which carries 2,000 passengers and many vehicles. More modern versions ply the waters of Puget Sound. In warmer waters, the Vaporettis of Venice and the riverine hydrofoil systems of the Soviet Union have demonstrated the capability of high-speed, high-volume, comfortable, and low-cost mass transportation from the very heart of the dense metropolis to the most remote unpopulated areas without additional capital investment in roads or fixed guideways.

The outlines of a sea-based solution are, however, inherent in current developments in the industrial uses of the sea. An evolutionary process which is progressively producing a seaward extension of industrial and urban systems is rapidly unfolding. The thread of progression of significance in developing a sea-based city complex is as follows: 1. the conceptualization, design, and construction of the dynamically-positioned semi-submersible platform, 2. the conceptualization and concept design of the pre-stressed concrete floating city module, 3. the conceptualization, design, and initial stages of construction of the barge-mounted offshore nuclear power plant, 4. the conceptualization, design, and construction of the pre-stressed concrete EKOFISK oil storage facility, 5. the symbolic demonstration of the floating city concept at the 1975 International Ocean Exposition at Okinawa, utilizing a semi-submersible as the demonstration platform for the AQUAPOLIS concept (Figure 3), 6. the conceptualization, design, and construction of the "verticle bottle" semi-submersible as employed in the CONDEEP oil storage tanks, 7. the conceptualization, design, and construction of the barge-mounted floating industrial plant, 8. the conceptualization and concept design of the barge-mounted and semi-submersible-mounted floating coal power plant (Figure 4), 9. the conceptualization and initiation of development of the ocean thermal energy conversion plant.

This set of developments makes clear that the economically feasible urban platform is within the current state-of-the-art.

These specific milestones are just a few of the total set of offshore platform developments which have been initially associated with the oil industry and are now increasingly associated with oil, energy, and industry. The offshore development was initiated by the development of the offshore fixed platform whose success was followed by a bewildering variety of drill ships, jack-up barges, and semi-submersibles. In deep water, fixed platforms or moored systems became progressively non-feasible. It was a discovery of the National Science Foundation Deep-Sea Drilling Project that dynamic positioning (the use of rotatable propellor systems to counteract drift) was feasible and required relatively little energy in most oceanic situations. This resulted in the design of a structure called the MOHOLE platform. Although never completed, it became the prototype for a large number of semi-submersible oil rigs. Such platforms like the semi-submerged ship consist of twin submarine hulls connected by vertical columns to a box-like structure located well above the free surface. Insofar as distribution of usable space is concerned, it matches that desired for the idealized three-dimensional city module. Although these platforms are far more stable than conventional ships, their commercial function has not required the freedom from sensible motion that would be required for urban uses. Similarly, the

Figure 3. AQUAPOLIS, the demonstration floating city at the 1975 International Ocean Exposition, Okinawa, Japan.

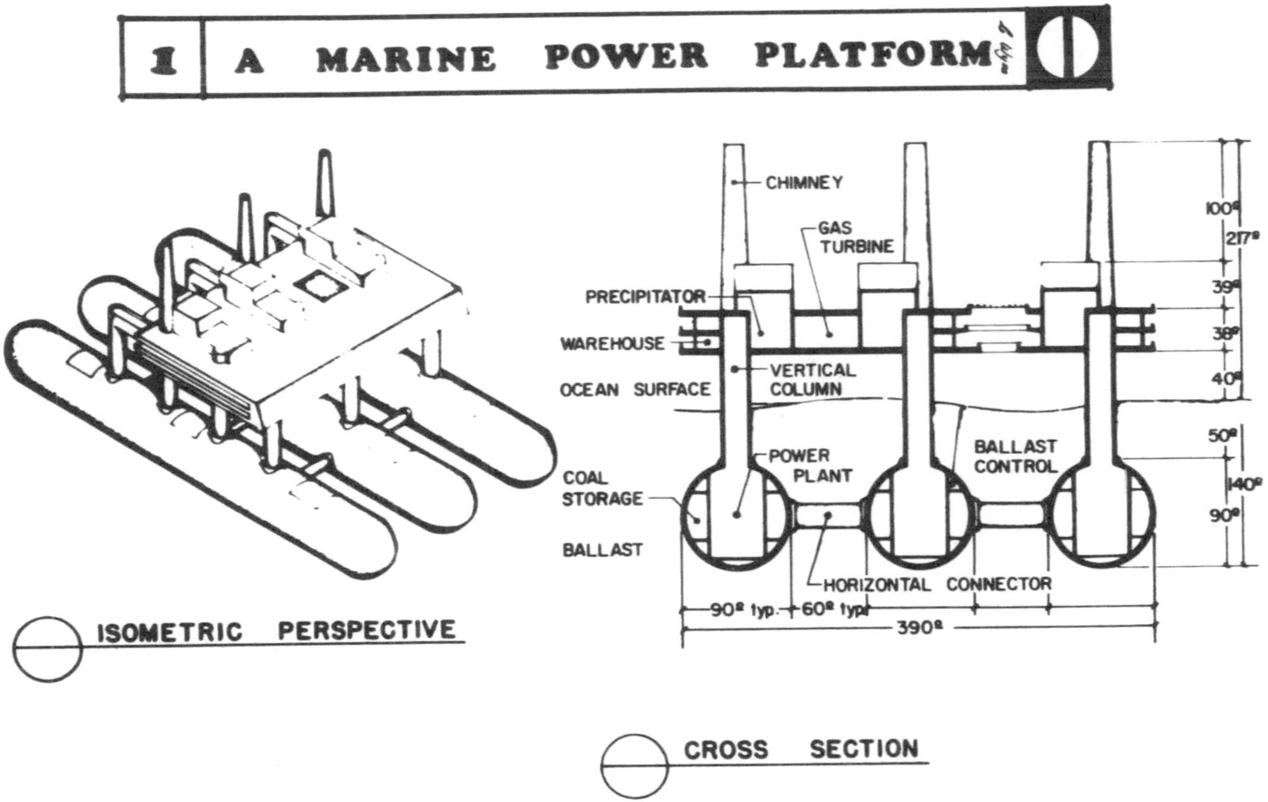

Figure 4. University of Hawaii floating coal-fired power plant.

volumes of usable space which are far greater than required for their commercial function are nonetheless required for urban purposes.

It was, however, the recognition of these characteristics that led to the basic design of the Hawaii floating city concept (Kikutake, Craven, et al.). In this conceptualization, the stable platform configuration consists of a triad of vertical column structures which are fully submerged. Each of these columns has a column of very much smaller diameter superimposed on top which penetrates the free surface, and supports a multi-deck structure located well above the free surface. Particular care was taken to find the theoretically most efficient structure for maximum stability in a seaway (Seidl) and, as a result, the required "motionless" configuration is obtained for platforms of large but nonetheless reasonable scale. At the time of this design, it was postulated that the structures would be of pre-stressed concrete although this technology had not yet been proved. It was also postulated that power plants and industrial systems for offshore platforms were economically and technologically feasible.

Contemporaneous with the floating city design was the conceptualization and design of the Westinghouse Corporation's floating nuclear power plant. These barge-mounted plants are to be located along the Eastern seaboard in the relatively shallow waters overlying the continental shelf. The power generated by these plants is to be carried by underwater cables to the user. Calculations indicated a large number of superiorities of these plants over their land-based counterparts as follows: 1. reduction in costs due to multiple production in shipyards, 2. reduction in costs and increase in safety due to elimination of structural foundations, 3. reduction in costs and increase in safety by elimination of hazard of earthquake, 4. increase in safety and efficiency of the city by the location of the power plant at a considerable distance from the user, 5. increased safety from sabotage by virtue of remoteness, 6. aesthetic and non-polluting by virtue of remoteness, 7. low in cost by virtue of sea-based supply, 8. protected against storm by a large dike, the plant is virtually immune to natural catastrophe.

On the basis of their study, Westinghouse Corporation has proceeded with the construction of a "shipyard" for the sole purpose of building the floating plants and with the contractual and permit process to build the first of these units. Unfortunately, the regulatory, socio-economic, and self-punitive characteristics of our current social and political organization have virtually prevented the completion of this project. It nonetheless demonstrated that floating nuclear power plants (whose technical feasibility was demonstrated in the nuclear submarine) are commercially feasible.

The conceptualization of the floating city was also contemporaneous with the conceptualization, design, and construction of the EKOFISK offshore oil storage facility. This major innovative development which revolutionized the use of concrete for offshore structures was constructed in Stavanger, Norway. The tank, designed to store one million barrels of oil, consists of nine nested cylinders approximately 80 feet in diameter and 280 feet tall (the height of a 20-story building). The cylinders are made of pre-stressed concrete. The entire assembly is surrounded by a concrete wall which is perforated in a manner that will most effectively absorb the energy of storm waves. The facility was built by the initial establishment of a dewatered cofferdam. The base was poured in this area and the vertical columns constructed by a slip form technique until they were tall enough to be floated. The cofferdam was then removed and the installation floated into deeper water. Construction was completed by continuation of the slip form technique while the structure was floating in deep water. It was then towed to the site and sunk in place. Although visited by a major fire, the EKOFISK facility has successfully demonstrated the ability to build large volume structures in sheltered bays and to tow them for use in the open ocean.

The Hawaii design pointed the way to a symbolic demonstration of the floating city concept of the 1975 International Ocean Exposition at Okinawa. Although contemplated, time precluded the construction of a dynamically positioned vertical column platform. A more conventional semi-submersible moored rig was chosen to represent the floating city module. Significant features included the dual location of the AQUAPOLIS either connected to the land by a fixed bridge or located at sea and accessible only to marine transportation. It could also be repositioned well at sea should a typhoon of extreme intensity be anticipated. A representative set of city services were provided, which included power generation (fossil fuel), waste treatment, fresh water production, etc. Another significant demonstration was the fact that the AQUAPOLIS was constructed at a shipyard in the main islands and located in Okinawa with a minimum of site preparation. The AQUAPOLIS did not have the requisite stability (i.e., visitors could sense slight motion of the platform), the volume of usable space, or the low cost per unit volume that would be required for an economic urban module.

Fortunately and once again contemporaneously with the development of AQUAPOLIS, was the conceptualization, design, and construction of the CONDEEP oil storage facilities. These pre-stressed concrete structures are of the size and configuration which were postulated as ideal for the floating city module. Although they are not dynamically positioned in their location as storage tanks, they were nonetheless built at a site remote from their final location and were effectively towed from the construction area to the site location. These situations demonstrate beyond the shadow of a doubt that massive pre-stressed concrete elements can be built in a size and configuration that can accomodate the volume and arrangements requirements of the ideal three-dimensional city complex.

Although the CONDEEPS go far in demonstrating the feasibility of structure, they do not demonstrate the feasibility of location of industrial facilities at sea. Two studies were conducted in Hawaii on the feasibility of location of conventional coal power plants. One study postulated a barge-mounted plant, the other a semi-submersible platform. Both indicated that the cost of such a plant would be the same or less than the cost of a land-based plant of similar capacity. The advantages of sea-basing which were discovered in the Westinghouse nuclear plant design were found to be equally applicable to the conventional power plant. In addition, and depending on plant siting with respect to prevailing winds, it was found that the polluting aspects of coal-fired plants could be substantially minimized.

These studies have been effectively confirmed by the actual construction and installation of barge-mounted industrial plants which have been built in Japan and installed in far distant overseas locations. Ishikawajima-Harima Heavy Industries has completed plans for barge-mounted steel mills, cement plants, thermal power plants, a utility plant, a refinery, a sugar plant, and a desalinization plant and has completed and installed a pulp plant with its associated power platform. The plants were built in Japan and installed in the Amazon. The journey was antipodal and thus demonstrated the global capabilities inherent in this form of construction.

The sum of these developments is such that it is now technically feasible to evolve or develop an urban community which matches our ideal model. This is illustrated by a configurational layout on a typical coastal area on the west coast of Japan. A small island offshore serves as the locale for the local airport. Industrial facilities, power generation facilities, and waste disposal facilities are located on separate and separated floating complexes. The urban and industrial center is located on a slowly mobile platform at the mouth of the valley. All of these are connected by marine mass transit to the interior along the streams of the valley. Farms, agriculture, schools, stadiums, government plazas are located in the lower valley, and in the upper valley, the temples, parks, and teahouses of an earlier era are preserved and made available. Note that little, if any, structures are located on the coastline and that the coastal zone itself is relatively uncluttered except for recreational uses. The concept fully recognizes that the coastline is the delicate and fragile ecosystem and that the land above the highest tide level is moderately robust and that the ocean beyond the bay and estuary is very robust. In general, the choice for location of each element of the urban system is between the land well within the valley or delta or the sea a mile or more from the coast with both complexes tied together by a transportation system which is orthogonal to the coastline and which penetrates to the uppermost reaches of the navigable streams.

If a case has been made that it is possible to build a large-scale urban complex that is economic and functionally beautiful in the sense that it preserves small-scale phenomena and satisfies the design criteria for the psychologically satisfying city, one may also ask the more trivial question as to whether such complexes can be aesthetic from an architectural sense.

Except for the AQUAPOLIS, the author is unaware that an architect in the classical sense has ever been employed in the design of any of the offshore structures now extant. (The preliminary refinery plant on an island in the harbor of Long Beach, California, is another exception. The plant camouflaged to look like highrises on a palm tree island can be deemed aesthetic or not, depending on your attitude toward highrises.) Two factors have obviated the need for such architectural treatment. The first is the fact that the major portion of the offshore structures are submerged and therefore not visible. Observers directly above can see with clarity no further than 50 feet in the clearest of water and even then the view is obscured and enhanced by the filter of blue. The second is the fact that platforms at a distance from shore soon fall below the horizon and are not seen except for those few who man the facility or visit for purposes of inspection.

An urban complex will, of necessity, be close enough to shore and will have enough of its structure above water to warrant architectural treatment. Such considerations were paramount to the Hawaii floating city design. A prime factor in aesthetic enjoyment was the mobility requirement which would have resulted in the complex moving back and forth (albeit very slowly) in a more or less prescribed track, and rotating. Thus, the view from shore and the view from sea would be constantly changing. A dweller in a peripheral apartment would have an ephemeral view--on some days, the view would be of sunrise, on others of sunset; on some days of land and on others of sea.

It should not be difficult to make the overall profile as seen from a distance quite attractive. Many artists have found the seascape to be a subject of great appeal and have found it convenient to employ floating objects as the centerpiece of the masterwork. In general, these views have been pyramidic with the water serving as the base of the pyramid, the structure serving as the lower base, the superstructure as the upper portion, and the sun or moon as the single illuminator located at the apex.

The Hawaii floating city configuration was designed with this concept in mind. Although it is true that most architectural models and drawings are beautiful whether or not the prototype proves to be so, the floating city model leaves little doubt of grace and beauty (Figure 5). The webbed curtains which are employed for screening sun, and as a windbreak, provide the graceful flowing surfaces which carry the eye from the horizontal surface of the water to the delicate minarets. No structure above the surface is more than eight stories tall, nor need it be, for the bulk of the usable volume is located below the surface. The total design, though big, is beautiful; it is functionally complete, a concept of land and sea, a thing of beauty, and a joy forever. It is to be hoped that someday the regulatory, socio-economic and self-gratification structure of the United States will permit the aggregation of capital and the unfettering of imaginative and competent architects and engineers so that this concept and others of its ilk can be developed for the betterment of man.

Figure 5. University of Hawaii floating city model.

Overview

American agriculture is on the wrong track. The farming methods of "agribusiness", the policies and programs of the federal government and the states, and the exaggerated use of chemical fertilizers are leading to massive erosion of topsoil and the destruction of the soil's biological resilience. The emphasis should shift to ecologically-sound farming methods and to the restoration of the family farm. James A. McHale argues that changes in the institutional policies and the technical methods of American agriculture are necessary to insure the nation's ability to produce abundant and healthful food in the future.

James A. McHale

11. The Future of Rural Technology

Superficially, the American agricultural system appears to be one of our great success stories. In an age when 500 million people go to bed hungry each night, the United States is able to feed its own populace (perhaps too well) and export 25 billion dollars worth of farm produce annually. Farming is just over 3% of the gross national product, but last year it produced 20% of all our exports. But that production was achieved through our past and present agricultural land and food policies which encouraged agribusiness at the expense of the farm family, the environment, consumers, and our soil. This is, I believe, one example of "big but not beautiful."

My thesis is that we desperately need a program that will recapitalize and revitalize Rural America and which will encourage ecologically sound farming methods, preserve threatened agricultural land, stimulate rural and small town development, and insure sound nutrition in foodstuffs for man and animal.

I want to tell you, "Things are not rosy down on the farm." There are fewer than 3 million farms today in the United States, half the number in 1950.[1] Farm debt is $136 billion and debt service takes 10% of all farm gross income. Farm income is at 70% of parity.[2] We have sent 30 million people to the cities in the last thirty years, and many have not found a better life there. Many of these displaced rural families have contributed to the increase in welfare payments, unemployment, and inner-city destruction. We still are losing rural population at a rate of 700,000 a year. If we continue to lose farm families at the present rate, we may well be minus the small farmer long before the year 2000.

We need the family farmer and the production philosophy he represents to maintain a quality of life for Rural America and to produce a better quality of food for all Americans. The economic pressures which squeeze the family farmers out result in corporations becoming increasingly involved in every phase of our food production and distribution system. (Fifty corporations now control 75% of all food manufacturing assets.)

Agribusiness as presently structured has not been an unqualified success for the rest of the U.S. population. Consumers face steadily increasing food prices and declines in the nutritional value of food. It has been estimated that the American diet contains 166 pounds of sugar and 9 pounds of food additives annually. Add this to the 5 pounds of herbicides, pesticides and fungicides that are manufactured each year for each man, woman and child in the United States, at least some of which are consumed in the final product, and I think the case is very strong for changing our direction in agriculture at once.[3] Indeed, there is some speculation among scientists and doctors that some of these chemicals, while not affecting our health in ways that we immediately notice, might cause genetic change and damage in future generations. There are other problems also: we all know the DDT -- 2,4-D -- 2,4,5-T -- PPB stories, but we don't know the final results. Let's hope Rachel Carson's "Silent Spring" doesn't come true.

I feel that we are racing against time. I can't visualize the United States remaining a viable nation unless we preserve and encourage relatively small and independent business entities, including the family farm. To do this we need a rational land policy, and the first step is to find out, "Who owns America -- Pennsylvania, Maine, Texas, California, and the rest?" In fact, twelve land owners own 52% of the land area in Maine, 25 own half of California; and we even see the United States government encouraging foreign investors to buy American farm land.[4]

The loss of our millions of farm families was

not by accident; rather, our cheap food and agricultural policies systematically favor the large corporate farms through tax breaks, credit provisions, and research geared to the "bigger is better" philosophy without regard to long-term impact on our environment, the farm family, or our rural culture.[5]

All this has been supported and nurtured by our U.S. Department of Agriculture, land-grant colleges, Extension Service, and some state departments of agriculture, in the name of progress, free enterprise, and supply and demand. The direct and indirect influence from agribusiness has been substantial if not controlling.

The Hatch Act of 1887 instructed our land-grant institutions to promote "a sound and prosperous agriculture and rural life," a permanent agriculture, and it says their purpose should be the development and improvement of the rural home and rural life. As far as I am concerned, the land-grant college experts have presided over the destruction of what they were mandated to preserve. They function in what has become a closed, self-serving society. I believe that the land-grant institutions have used the wrong standards. They have used industrial standards concerned only with speed and efficiency and have applied them as standards for agriculture. These are short-term concerns and even in industry they are not the only appropriate concerns. In agriculture where we are working with renewable resources such as soil that must be maintained and conserved, such standards are clearly inadequate.[6]

We should have learned from history. We have seen nations fall when they lost their renewable resources of topsoil and forest land. We have seen the Sahara Desert moving south at the rate of several miles a year after the forests were destroyed and the eco-balance lost.

We in the United States have been able to destroy a lot of our land in fifty years. There are those among us who remember the dust bowl days of the Thirties, when Oklahoma blew over New York City, and we are <u>still</u> losing our top soil. Through water erosion alone, our top soil is being drained at a staggering rate: nationally, erosion averages nine tons per year, more than twice as fast as the soil can possibly be replaced. This is occurring not only on our hillsides, but also on relatively flat, rich farmland where the American farmers are following intensive planting methods without the requisite intensive soil preservation practices.

Let me quote the following excerpts from a Government Accounting Office report to the Congress -- CED 77-30, February, 1977:

> The Council for Agriculture Science and Technology reported in January, 1975, that more than 1/3 of the United States cropland was suffering annual soil losses in excess of the limit at which soil productivity can be sustained over time. The report suggests that we are less effective in controlling erosion today on some lands with severe erosion problems than we were 15 years ago.[7]

An April 1972 Iowa State University research report[7] stated that the United States was losing 4 billion tons of top soil a year through water erosion, compared to 3 billion tons in 1934. The report said that it would take a train of freight cars about 633,000 miles long to move 4 billion tons of top soil--a train long enough to circle the earth 24 times! The report also said that today's farmers were losing an average of 12 tons per acre annually through water erosion, as compared to 8 tons in 1934. When I am talking about a ton of top soil, I am talking about a layer the thickness of one sheet of typing paper, spread over an area of one acre. Another regional example of erosion is that the equivalent of a 120-acre farm floats down the mouth of the Mississippi River alone each day.

In June 1974, the Soil Conservation Service reported that Iowa had experienced its worst spring for soil erosion in 25 years. It stated that many fields had soil losses of 40-50 tons an acre and that some steep slopes lost as much as 200 tons an acre.

As you can see, the extent of soil erosion in America is massive, but the loss of soil nutrients by erosion is even worse than the volume of erosion would indicate. In New Jersey, for instance, soil materials removed from test plots by the selective action of erosion contained four to seven times as much organic matter, three times as much phosphorus and 1.4 times as much potassium as the original soil in the plots.[8]

The GAO report concluded that the rates of estimated soil losses revealed by their review, in combination with other reports, indicate that demise of the nation's agricultural productivity potential is possible.

And what has been done with the report? As far as I can see, it has been tucked away with the mountains of other GAO reports. It takes guts, imagination and work to make things happen, and we seem to act only in times of crisis. I am sure that when America wakes up to the fact that our lives depend on that few inches of top soil, we will respond. However, I am also sure President Carter and his budget-cutters will propose one more time to cut the Agricultural Conservation Program budget from the USDA budget. FDR had foresight and concern for the American farmer and our soil. His budget had $500 million for ACP--cost sharing--for farmers in the depression years of the Thirties, for conservation measures. Now ACP has been a mere $190 million of inflated dollars for several years running.

Clearly, we need a new approach to soil conservation. We have oil depletion allowances to encourage oil production. We also need a soil depletion allowance for our farmers to be used for long-term conservation practices to rebuild out land. I think a Des Moines reporter summed it up well a few months ago when he said, "Economics and politics have combined to block any real solutions to these environmental problems associated with American agriculture."[9]

I believe the evidence is there that federal and state agencies have mismanaged our soil conservation programs; Congressmen have invoked parochial concerns to block effective curbs on farm-caused pollution; and the farm organizations themselves have been complacent and even hostile to government attempts to deal with the situation.

It was Secretary of Agriculture Earl Butz who ended our land conservation policies in the early 1970's and urged our farmers to plant fence-row to fence-row, then to tear up their fence-rows and shelter belts and plant those areas too. This increased production helped the Russians in the famous Russian Wheat Deal. It also helped our balance of trade deficit. It improved farm income in the short-run--we sold the grain--but now our fragile, abused soils are washing down our slopes into our already polluted rivers.

Our present system of macro-agriculture is ruining our soils in another way. By encouraging the increasingly heavy use of commercial fertilizers, notably nitrogen, the soil biota vital to fertile soils are being destroyed.[10] Agriculture still depends heavily on nitrogen fixation, but our present agriculture practices using anhydrous ammonia fertilizer are destroying this possibility. Only a few spokesmen in our agriculture research and educational establishments are aware of or are concerned about the decline of the organic matter in our soils.[11]

Unfortunately, we have no reliable current information on the fertility of American agricultural lands. According to the GAO report cited above, the priorities for money use were misguided, and projects that increase production at the expense of conservation practices were funded. Also, their survey of the Soil Conservation Service functions pointed out that much of the technical information actually gathered is unwanted and unused by the farmers. What is needed is an outreach program to inform and educate the farmers on real, practical, workable soil conservation practices. And certainly, it is necessary to coordinate the efforts of the different USDA agencies, so that all are going in the same direction; instead of the county agents, Agricultural Stabilization and Conservation Service, and the Soil Conservation Service all coming up with different and conflicting recommendations.

Let me be more specific about what USDA has not done:

1. USDA has not collected data on the humic soil levels for some 30 years.

2. Most research on organic levels of our soils is over 10 years old.

3. No one has been working specifically on the question of numbers of soil microflora, and there is no reliable information on the national impact of chemicals and monoculture farming on beneficial soil biota levels.

4. Methodologies for determining soil tilth are non-existent.

I, for one, must ask, aren't the agencies concerned about the above questions? Or do they feel that our soil is only for holding a seed and plant in place, to be chemically injected, not naturally nurtured?

Our agricultural research establishment has successfully contributed to high production but has ignored the soil, the basis of that production. The future of soil productivity is threatened, and we abound in ignorance concerning the second-order effects of the miracle chemicals, particularly anhydrous ammonia. We appear equally ignorant of the causes for reduced nutritional value of our food.[12] The current increase in pest infestation and crop disease is also a serious problem. We have raised hybrid bugs so strong that we can't kill them between two flat boards. Everything is out of balance.

An acre of fertile topsoil contains eleven tons of biological life. Modern farms using toxic technology often have less than two tons. Fertile soils have as many as five million earthworms per acre. Modern farms using acid fertilizer have less than 100,000 worms per acre. The natural ability of the soil and its organisms to decompose organic matter has been substantially impaired. Farmers are finding cornstalks that they plowed under three years earlier <u>still</u> showing no signs of decomposition. Our soils are suffering from thirty years of toxic chemicals upsetting the soil's eco-system and our total disregard of the long-term effects of these chemicals on soil, plant, animal and human life.

Until recently the effects of soil erosion and

soil deterioration on crop production have been masked by the increased amounts of fertilizer being applied to the land.

But the mask is slipping:

1. While consumption of commercial nitrogen doubled from 1966 to 1977, U.S. farm production increased by only 23%, 10% of which came from cropping more land.[13]

2. It is estimated that it takes five times as much fertilizer today to produce the same yields as it did in 1949.[14]

3. Yields per acre of wheat, maize, sorgum, soybeans, and potatoes have not increased since 1970.

This has happened despite the increase in chemical fertilizer usage--20% more was used in 1976 than in 1975.

If change is to come, we need a national commitment and support from our government and educational institutions to conserve our most finite resource--our soil. In 1978, land-grant institutions devoted 6,000 man-years to production efficiency projects and only 258 man-years to rural development. I think the priorities are mixed-up and would like to make a few recommendations so we can begin to turn things around:

1 Congress should require the USDA to use 50% of its research funds for small farm demonstration and research projects geared to soil and energy conservation projects.

2. A substantial amount of agricultural research funds should be ear-marked for testing new ideas and products. There is little support from our public institutions for change. There are 4,000 experts on pesticides in Iowa alone, yet I don't know any land-grant institution that teaches organic farming or the ecological approach to agriculture.

3. The testing of new products and concepts should be run "in the open" by the land-grant schools and extension services, with free access for the public to view the action, rather than being conducted in a shroud of secrecy.

4. I would urge the Department of Energy to establish a national, low-energy, demonstration farm and rural life center where new products, innovations and concepts can be presented and tested. Further, I want the developers of these ideas to do the testing, openly of course, and under supervision, to ensure unbiased testing.

5. A major soil improvement effort is needed, and this will imply institutional innovations such as a permanent civilian conservation corps.

There <u>are</u> limits to growth. I am here to tell you things are going to change whether we like it or not. We cannot continue to use energy- and chemical-intensive agricultural methods. Moreover, half of the world's land available for agriculture will be consumed by urban and industrial development by the year 2050 and population may quadruple. There is an immediate need to support innovative agricultural research in this country, to meet the winds and tides of change, for the survival of humanity, and to restore and maintain the soils on which our food is grown.

References

1. Perelman, Michael J., Farming with Petroleum, "Environment", October 1972, Vol. 14, No. 8, p. 8.

2. USDA Agricultural Handbook No. 551.

3. Pimentel, David, J. Krummel, D. Gallahan, J. Hough, A. Merrill, I. Schreiner, P. Vittum, F. Koziol, E. Back, D. Yen and S. Fiance, Benefits and Costs of Pesticide Use in U.S. Food Production, "BioScience", December 1978, Vol. 28, No. 12, p. 772.

4. Weiman, Dave, National Farmers' Union Director of Research, Washington, D.C., Taken from speech "Land, Agriculture and People", prepared April 27, 1976

5. Hightower, Jim, Hard Tomatoes, Hard Times, Task Force on Land Grant College Complex; Agribusiness Accountability Project, Washington, D.C., 1972.

6. Brown, Lester, Historic Exploitation of Cropland, "Worldwatch Paper #24 - The Worldwide Loss of Cropland", October, 1978, pp 7-8.

7. "To Protect Tomorrow's Food Supply, Soil Conservation Needs Priority Attention", Dept. of Agriculture, Report to the Congress by the Comptroller General of the U.S. - General Accounting Office (GAO), February 14, 1977, CED-77-30.

8. World Crisis in Agriculture, The Ambassador College, Agricultural Research Department, Ambassador College Press, Pasadena, CA, 1974.

9. James Risser, Washington Bureau Chief for the Des Moines Register, September 15, 1978.

10. Belanger, Jerry D., Soil Fertility, Countryside Publishing, Ltd., Waterloo, WI, 1977.

11. Zweidling, Daniel, Can U.S.Farmers Kick the Petrochemical Habit, "New Times", May 29, 1978, pp 23-65.

12. World Crisis in Agriculture, The Ambassador College, Agricultural Research Department, Ambassador College Press, Pasadena, CA, 1974, pp 19-25.

13. USDA Staff Bulletin #581.

14. Senator James Abourzek, then Chairman of Senate Judiciary Subcommittee on Deficiencies in Agricultural Research, Transcript of Hearings held in October, 1977.

Suggested Readings

1. Acres, U.S.A. Newspaper, Published Monthly, 10227 E. 61st St., Raytown, MO 64133.

2. Albrecht, William A., Soil Fertility and Animal Health, Fred Hahane Publishers, 1958.

3. Balfour, Lady Eve, The Living Soil, Faber Pub. Co., 1943.

4. Belanger, Jerome D., Soil Fertility, Countryside Publications, Ltd., 1977.

5. Berger, Samuel R., Dollar Harvest, D.C. Heath and Company, 1971.

6. Blake, Michael, Concentrated Incomplete Fertilizers, Crosby Lockwood Co., 1966.

7. Bromfield, Louis, From My Experience, Harper Pub. Co., 1955.

8. Countryside Magazine, Published monthly, 312 Portland Rd., Waterloo, WI 53594.

9. Hightower, Jim, Hard Tomatoes, Hard Times: The Failure of the Land Grant College Complex, Agribusiness Accountability Project, 1972.

10. Mother Earth News Magazine, Published monthly, 105 Stoney Mountain Rd., Hendersonville, NC 28739.

11. Pimentel, David and Krummel, John, America's Agricultural Future, "The Ecologist", Vol. 7, No. 7, pp 254-261.

12. Pimentel, David, et.al., Benefits and Costs of Pesticide Use in U.S. Food Production, "BioScience", December 1978, Vol. 28, No. 12, pp 772-784.

13. Pimentel, David, et.al., Energy and Land Constraints in Food Protein Production, "Science", November 21, 1975, Vol. 190, pp. 754-761.

14. Pimentel, David, et.al., Food Production and the Energy Crisis, "Science", November 2, 1973, Vol.182, pp 443-449.

15. Pimentel, David, et.al., Land Degradation: Effects on Food and Energy Resources, "Science", October 8, 1976, Vol. 194, pp. 149-155.

Section B

Perspectives on Transportation

The Seneca Chief, the first canal boat through the Erie Canal, passing through the deep cut at Lockport, New York, October 1825. Reprinted with permission from Cadwallader D. Colden, Memoir...At the Celebration of the Completion of the New York Canals (New York: The Corporation of New York, 1825). Courtesy of the Canal Museum, Syracuse, New York.

Overview

Construction of the Erie Canal illustrates the benefits of macro-engineering in the less complicated--and less congested--world of the early nineteenth century. Jeanne Krause shows how a relatively small investment produced profits for investors, better terms of trade for users, a healthier economy for a whole region, and an example of inter-sectoral cooperation on the grand scale. The canal was a major contributor to the "take-off stage" in national economic growth. But despite the logic of its appeal, the sums needed for the canal seemed exorbitant at the time: the whole project might have languished had it not been for the astute and persistent promotional efforts of DeWitt Clinton.

Jeanne Krause

12. The Erie Canal: Macro-Engineering When the World Was Still Simple

My paper on the Erie Canal might properly carry another subtitle: rarely before have so few done so much for so many with so little. With my best apology to Sir Winston, this happened 125 years earlier than the R.A.F.'s defense of Britain.

The Erie Canal joining the Hudson River with Lake Erie was the remarkable achievement of determined men and the preeminent product of the rapid economic growth in the first decades of our republic. It was the product, too, of the close collaboration between New York officials and private entrepreneurs bound together by strong economic ties. In just seven years -- from 1819 to 1825 -- they gouged a ribbon of water 363 miles through swamp and rock, over rivers, past Indian settlements and wilderness to link New York City, "the emporium of America," with the extensive, remote frontier.[1]

This ribbon of water captured the imagination of New Yorkers in the early days of the 1800's, and naturally found its way into literature and folklore. The canal was called "The Big Ditch," "The Grand Canal," and "Clinton's Folly." Towns not on the ocean or rivers earned names such as Newport, Middleport, Lockport. For decades afterwards, when someone in a crowded room shouted, "Low bridge!" -- everyone else ducked.

---

[1]De Witt Clinton, Memoir of DeWitt Clinton, Reprinted by David Hosack (New York, 1829). "Ribbon of Water," Shaw, p. 87.

This ribbon did all that was expected of it, and more. DeWitt Clinton prophesied that the canal would promote national union and national defense, reduce transportation costs, deflect commerce from Canada to New York, and speed settlement of western U.S. territories.[2] This the canal did. In the process, it also helped guarantee New York's emergence as the principal city of the new nation. This extraordinary enterprise was, in Nathan Miller's phrase, "The Enterprise of a Free People." It was also a major contributor to what Walt Rostow called the "take-off" stage of national economic growth in the 1840's.[3]

The canal was the <u>only</u> enterprise of a people, bent on building canals in the first half of the 19th Century, that truly succeeded. Between 1815 and 1860, 4,254 miles of canals were built in 23 states for an estimated cost of $188 million. About 90 percent of that cost was financed by government bonds, half of which were ultimately held by foreigners, especially the British. The Erie Canal provided the greatest benefits and was the longest in service -- until 1910, when the current Barge Canal was built.

In part, the Erie Canal's success was due to the relative economic stability of the 1820's, when money markets functioned smoothly. And 50 years of peace were just beginning. In part, its success came because it was the beneficiary of historic and geologic coincidences, and of extraordinary luck.

New York state had a mercantilist tradition that led them to accept state aid, "regulation, and intervention in the economy." This "entrepreneurial role" of the state led to close economic ties with its citizens.

---

[2]DeWitt Clinton, "Memorial of the Citizens of New York, in favor of a canal navigation between the great western and northern lakes and the tidewaters of the Hudson." In <u>Laws of the State of New York in Relation to the Erie Canal</u> (Albany, 1816), pp.122-145.

[3]Walter W. Rostow, <u>The Stages of Economic Growth: a Non-Communist Manifesto</u> (Harvard University Press, 1960) p.38.

Two wars -- the Revolution and the War of 1812 -- had proved the military necessity of cheap transportation for defense of the frontier. For example, in 1812, a $400 cannon cost $2000 to move across New York State.

Finally the Mohawk River valley was the only major east-west break in the Appalachian chain from Georgia to Maine -- mountains which pinned the population on a thin strip of Atlantic coastline, mountains which kept abundant natural resources beyond their reach.

The bare facts were extraordinary enough. When it opened in 1825, it had 83 locks raising the water level 565 feet above the Hudson River, but lower than Lake Erie, a constant source of water. Eighteen aqueducts--marvels of the day--spanned valleys and rivers, including the Mohawk, twice. Although the canal was 363 miles long, it was only 40 feet wide and 40 feet deep. It cost $7,163,000; yet it was built by a state whose annual budget was only $700,000. For the cost estimators among you, the cost overrun was a mere one percent, and it was completed in half the anticipated time.[4]

This fragile ribbon of water was built by farmers, carpenters, surveyors, judges using rudimentary tools and some luck, DuPont blasting powder had just become available. New mechanical devices would be devised for felling trees without an axe or saw, and for grubbing 30 to 40 stumps a day. And by a twist of fate, the first segment to be built in the middle of the state passed very near an abundant supply of hydraulic cement, hitherto unknown in America.

In the early 1800's there were no engineers in America, they had to be imported. There was virtually no engineering training available anywhere until the U.S. Military Academy began in the 1820's. The Erie Canal became the school for engineers. In 1837, there were only 84 known engineers in the U.S. -- 19 from the academy and 11 from the Canal. "Graduates" of the canal would go on to build many of the other canals.

---

[4]George E. Condon, Stars in the Water: the Story of the Erie Canal (Doubleday and Co., 1974) p. 47.

Contracts were let in small segments, some as small as three-quarters of a mile. In fact, 50 men contracted for the first 58 miles in what was really a series of mini-canals, each one tested with water before settlement into the system. These contractors were mainly from adjacent areas, they furnished their own tools, hired their own labor. They used specifications drawn up by carpenters and surveyors during construction of early segments. The workers generally came from adjacent areas and, contrary to prevailing opinion, over three-quarters of them were American-born.

Most materials were found locally. "Blue mud of the meadows" provided canal lining to prevent seepage. A high grade limestone found near Medina was excellent for locks. And another variety, "meagere limestone," was the water cement mentioned earlier. Governor Clinton had read about Roman cement, and had sent a young surveyor to Europe to study canals. This enterprising young man spent his own money exploring canals, aqueducts, and experimenting with underwater cement. In 1818, he found a type easy to prepare, which hardened under water -- and he found it just miles from the first segment of canal to be constructed.

The first segment of 93 miles was built to connect the Mohawk Valley with the western counties. This choice of the middle for initial construction proved to be fortunate. Choice of the eastern section would have only duplicated an existing, if occasionally impassable, route over the Mohawk River valley. Choice of the western section would have produced a canal linking wilderness with wilderness. The middle segment, in effect, extended the potential for travel west from Utica, generated new trade in timber and grain, and passed through the salt flats on which Syracuse was built. The revenues from travel and taxes on salt and commerce were immediate proof of the canal's effectiveness--and guaranteed that the remainder would be built.

The engineering accomplishments still inspire awe. Aqueducts and culverts strode across obstacles on Romanesque arches. One embankment rose 70 feet above the Irondequoit valley.

The longest aqueduct, near Schenectady, was 1188 feet long and crossed the Mohawk River on 26 stone piers. Another, just a bit farther west, rose above the rapids at Little Falls, that had always required a portage for Mohawk River traffic. At that point, the Mohawk fell 39 feet in three-quarters of a mile. And in the middle of the rapids, the central arch of 70 feet rested on two piers driven down to bed rock.

Altogether, 27 of the locks were in the 30 miles between Schenectady and Albany. North of Albany, at the extreme eastern end of the canal route, stood Cohoes Falls -- always the most difficult geologic barrier to travel. There the river fell 400 feet into the Hudson River, and 19 locks were required to bypass the falls.

In the west, the canal pierced two miles of solid rock at Lockport, where a double flight of five locks carried the canal 66 feet over a rock ridge. Those locks required 50,000 feet of facing stone, were each 90 feet long and 15 feet wide, and cost $1,000 per foot of rise. Floors of the locks rested on timbers one-foot thick, which, in turn, were covered by two layers of well-jointed planks three and two inches thick. All timbers were thoroughly grouted and clamped in iron. And Lockport did not exist until July of 1821, when construction began.

For those investors among you, several years before 1825, revenues from the still incomplete canal exceeded interest costs; from the time the canal opened, tolls were double the debt interest. Moreover, surplus revenues from the canal, which the state reinvested, themselves earned interest. Seven years after its completion, revenues had surpassed the entire cost of construction. Tolls were abolished in 1883. By then, all revenues had totalled $122 million. All construction, supervision and maintenance costs had come to only $80 million. Construction, that is, of two canals-for New York decided to double the size of the canal just 10 years after it opened.

If you actually had been an investor in canal fund bonds, you or your heirs would have reaped another, unanticipated benefit. The commissioners, unable to retire the debt early, sought alternative investments for surplus revenues. Among their choices: newly-formed companies building what we know as the New York Central railroad system.

Funds for construction of the canal and revenues from its operation contributed to the economic development of regions through which it passed. Construction was paid for by engineers, using drafts on local banks. As revenues from passengers, salt, and goods poured in, they too were deposited in local banks; in 1826, just one year after opening the canal, the canal commissioners, unable to retire the debt -- then selling about 25 percent above par -- began depositing millions of dollars in New York City and banks along the route, at 5 percent interest. After 1831, the canal fund, in effect, became a development bank, as commissioners also loaned money to the banks. By 1836, they had loaned $3.6 million to 52 institutions. In the 1837 panic, the commissioners loaned $2.5 million in canal stock to seven New York City banks.

As one might imagine, shipping costs were drastically reduced by the canal, better for commerce than mud-clogged roads and unpredictable, flood-swept rivers. Before the canal, it took three weeks to move one ton of goods from Buffalo to New York -- at a cost of $100. The canal cut time to nine days and the cost to $10. More remarkable, the 90 percent reduction in costs wrought by the canal was more drastic than any subsequent differential brought by the railroads, and it allowed Americans for the first time to make effective use of the great interior and to establish the national market on which later industrial development could be based.

Population, of course, grew rapidly. To cite but two cities: in 1820 Rochester had only 1500 people, by 1850 over 36,000. Between 1820 and 1840, the number of people employed in manufacturing almost trebled, and those engaged in commerce and navigation rose tenfold. By 1835 Rochester had become <u>the</u> milling center of the state, its 21 mills using 96 stones. The city served the farmers who had planted wheat fields 40 miles deep on either side of the canal as it passed through Genessee valley. In 1830, 13 percent of all tolls on the canal were paid in Rochester.

Buffalo swelled from a hamlet of 2000 in 1820, to over 20,000 in 1835 -- the year $1 million worth of buildings went up. By 1850, the population had doubled again. The same occurred in other cities along the route and those that were to be founded in the west.

Land values were naturally affected. The Federalists' interest in the canal followed their interest in land. Many state officials, legislators, and canal commissioners promoting the canal, and later responsible for building it, were speculators in land. They were not disappointed. In one wilderness county crossed by the ultimate route, land values quadrupled three times: after the canal land survey was authorized in 1808, after the canal act of 1812, and again after the final act passed in 1817. They almost doubled again by 1819 when the first 83 miles opened nearby. What happened to those land values in the years in between, speculators know only too well.[5]

Politically, the canal had as great an impact as it had economically -- and was as subject to accident. As long as the idea was in the hands of visionaries, it made little headway in Albany or Washington, where early presidents were advocating using federal surpluses for internal improvements. Visionaries who were, by the way, wealthy Federalists owning large tracts of land and western New York legislators -- those who stood to gain the most from a canal.

The idea made little real headway for a generation, since the 1790's when an earlier attempt at a canal foundered because of bad management and under-capitalization, and its two miles actually built had wooden locks that were rotten almost before the canal opened. That abortive attempt did influence the New Yorkers' thinking about internal improvements. Failure to improve rivers strengthened the case for an independent canal, and financial failure of a private company led promoters to favor a public venture.[6]

---

[5]Jeanne Krause, unpublished master's thesis, "Speculation in Land Along the Route of the Erie Canal by New York State Officials 1793-1822" (University of Chicago, 1975).

Then, almost simultaneously, the idea of a canal got the support of New York City merchants and DeWitt Clinton -- a great, imaginative leader who organized the coalition that voted in the canal. Not a machine politician, Clinton had been drummed out of office as mayor of New York. Twice. It was easy in those days. After the second time, in 1815, he took up the issue of a canal, mobilized opinion throughout the state, and rose to the state house on a wave of sentiment for it. The coalition which passed the final act in 1817 was made up of Federalists and his splinter "Clintonians." Only two Federalists voted against the final survey enacted in 1816, and all voted for the 1817 act.

Clinton was masterful in overcoming opposition. To legislators and landowners from the Hudson River, he promised that New York City's population would grow at least as fast as the supply of competing goods from western New York. It did. To northern and southern counties not crossed by the Erie Canal, he promised a Champlain canal to link New York City with the St. Lawrence River and lateral canals to the Erie. They were built. In the process of forging a coalition of Federalists and western legislators, Clinton set in motion a process that would, even before the canal opened, move New York politics out of the hands of "the few the rich, and the well-born" into more democratic patterns. A move -- in the most populous state -- some scholars say was essential to the election of Andrew Jackson.[7]

Two other accidents are worthy of note. In the final days of legislative argument in 1817, people were still uneasy about the financial success of the canal until Martin Van Buren -- an early opponent of the canal -- tied his political future to the canal. We all know what his political future was. In the final days of argument, Van Buren proposed that canal fund bonds be backed by the credit of New York State. At that

---

[6]Nathan Miller, "Private Enterprise in Inland Navigation," in New York History, Vol XXXI, No. 4, October 1950, pp. 398-413.

[7]"The Few, the Rich, and the Well-Born," is chapter one in Fox.

time, private capital was in chronic short supply in America, and no state, including New York, was willing to raise taxes to build canals. They, therefore, had to rely on borrowing -- made much easier with the financial backing of the state. That proposal proved crucial to the act's passage.

In those days, the council of revision -- not the governor -- held power of veto, and the vote on the 1817 act was three to two against. Until former governor, Daniel Tompkins -- then vice president of the United States -- walked into the chamber and joined the debate. His opinion, he said, was that renewed hostilities with Great Britain were inevitable and that funds should be saved for war. Wherupon Chancellor Kent responded that he preferred a canal to war and voted yes. The act carried three to two.

The effort related to the canal contained certain ironies. In the early decades of the new republic, a tension existed between New England and the government in Washington. Federal jurisdiction over interstate commerce was not yet firmly established. The Erie Canal did more to force the government into regulation of that commerce than anything prior to the railroads. The New York Federalists worried about a possible shift of the national center of gravity away from New York to the Potomac and, under Jefferson, away from commerce and industry to agriculture. However, being in favor of a strong central government, they naturally turned to Washington for support and funds. Refused by three presidents and as many Congresses, New York State built the canal itself --thereby delaying the shift of commercial and financial power further south, but also, in a very real sense, tying together a fragmented nation.

Now, to us, it seems inevitable that the canal would have been built and its success assured. To citizens of New York State in the early 1800's, the scale of the endeavor appeared phenomenal, the cost exhorbitant. Technology was not yet adequate. Native engineers did not exist. To fully grasp what this enterprise meant to those citizens, let me suggest a comparable enterprise today: it would be as if Americans were to build a rapid transit system to the moon, without physicists or computers at a cost of $5 trillion -- to be borrowed mainly from the Russians and Chinese.

Would we calculate the costs and the benefits accurately? Would we accept the challenge?

Bibliography

Fox, Dixon Ryan, *The Decline of Aristocracy in the Politics of New York* (Columbia University Press, 1918).

Goodrich, Carter, editor, *Canals and American Economic Development* (Columbia University Press, 1961). Especially the chapters by Julius Rubin, "An Innovating Public Improvement: The Erie Canal," and by Harvey H. Segal, "Cycles of Canal Construction."

Miller, Nathan, *The Enterprise of a Free People: Aspects of Economic Development in New York State During the Canal Period 1792-1838*. (Cornell University Press, 1962).

Shaw, Ronald E., *Erie Water West: a History of the Erie Canal 1792-1854* (University of Kentucky Press, 1966).

Whitford, Noble E., *History of the Canal System of the State of New York*, a Supplement to the Annual Report of the State Engineer and Surveyor for the Fiscal Year Ending September 30, 1905, two vols., (Albany).

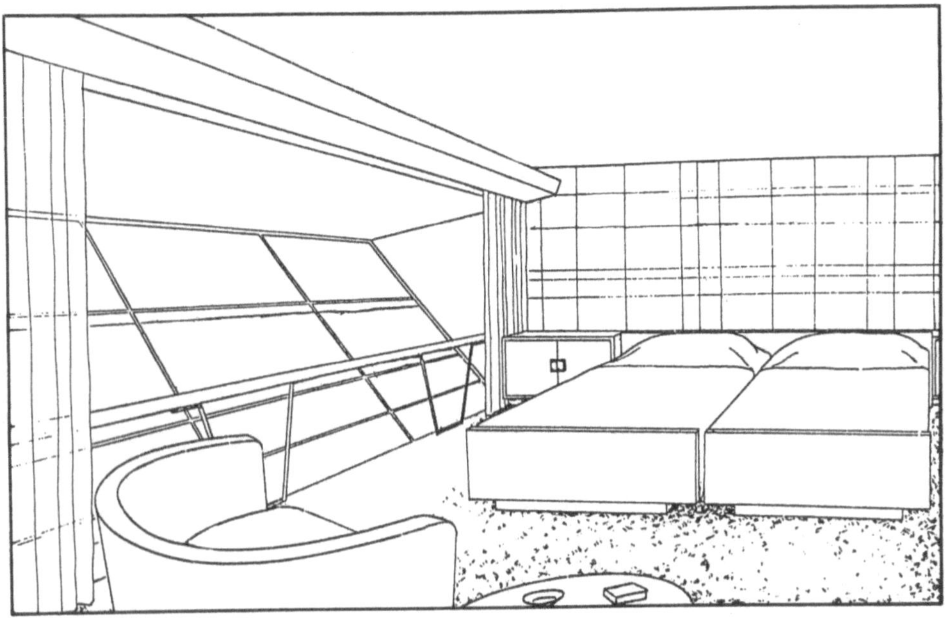

Conceptual drawings by Francis Morse of a "Deluxe Cabin" (above) and a "Sky Room" (below) in a "new generation" airship.

Overview

A new generation of giant lighter-than-air craft could provide a valuable type of air transport and find a viable economic niche in the spectrum of the transport industry. Francis T. Morse explains that structural efficiency can be enhanced by an increase in size to a point where an airship's ratio of useful lift to gross lift exceeds that of its heavier-than-air counterpart. Airships could utilize engines and fuels not suitable for heavier-than-air craft, one such possibility being a turbine system using liquid hydrogen for fuel. Among potential benefits of a new generation of lighter-than-air craft would be low-cost bulk transport and reduced air and noise pollution.

Francis Morse

13. The Future of Large Airships

The demands of long-range transportation have spawned an era of gigantism in both surface craft and aircraft. Notwithstanding certain disadvantages, the trend should generally not be unfavorably regarded, although it must be noted that its chief beneficiary is operational economy rather than structural efficiency. However, in the case of lighter-than-air craft, structural efficiency--a prime criterion for successful design -- can indeed be enhanced by increase in size, to a point where an airship's ratio of useful lift to gross lift far exceeds that of its heavier-than-air counterpart. This interesting statistic may prove compelling in spurring the renewed development of large lighter-than-air vehicles. But another factor will increasingly mitigate in favor of the airship -- the impending depletion of petroleum reserves to a point where alternative fuels must be introduced. Again, the giant airship casts its shadow upon the transportation picture.

It would be misleading to picture airships as playing the same role in aviation as airplanes; the inherent characteristics of each type are too divergent. The modern airliner has realized the role of the airplane in superlative fashion. But it has left gaps in the transportation spectrum which remind us of the potential value of the airship's role -- a long neglected one! If an airliner's principal asset is its speed, an airship's are surely comfort and luxury. The cruising speed of a passenger airship may never exceed 125 mph., but its potential for restful and luxurious travel is almost unbounded. Similarly, in the realm of air cargo, while the jet freighter can ferry big loads across the oceans overnight, the mean density of such cargo must be high because cargo space is limited. In an airship, weight - not space - is the only limiting factor, and consequently low density cargo can be transported effectively on intercontinental routes; if not overnight, in 36 hours at most, and possibly at lower DOC's per ton-mile.

With the above considerations in mind, let's examine the features and performance which will characterize a new generation of rigid airships. The hull of an airship is intrinsically a highly efficient space frame. With a fineness ratio of between 5 and 6 to 1, it is also efficient aerodynamically. Its drag is almost wholly skin friction. Past airships have not always been built with efficient hull structure, nor have they always been ideal aerodynamically - but this has been because designers failed to maximize their opportunities. Unlike heavier-than-air craft, the airship benefits greatly from increased size. This is because of the cube-square law which influences the ratio of dead weight to gross lift. I.e., since the hull volume (lift) increases as the cube of a linear dimension and the surface area (a fair measure of structural weight) as only the square of a dimension, the larger the ship, the smaller becomes weight/lift ratio. This fact should become a strong incentive to larger sizes in future airships. The Hindenburg and its sister ship represented the largest LTA vehicles ever built - each having hydrogen capacities of 7,000,000 cu. ft. with corresponding gross lifts of 500,000 pounds. Since existing hangars will, as a practical expedient, probably be used for the next rigid airship, a size limitation of some 12,500,000 cubic feet is indicated. Inflated with noninflammable helium its gross lift will average nearly 800,000 pounds. (more in winter, less in summer). It is sheer coincidence that this equals the takeoff weight of a Boeing 747, but it provides a welcome basis for comparison between the two types. Like the airliner, the airship will carry some 400 passengers. Unlike the jumbo jet, the airship will have deck space totalling some 40,000 square feet -- approximately 100 square feet per passenger, or ten times that of the airliner. The accommodations will, in fact, much more resemble those of a surface ship than an airplane. There will be staterooms, promenades, lounges, cinemas, saunas, gymnasiums and other novel amenities.

Structurally, it appears that for the size hull in question, the optimum structural arrangement should pretty much follow past practice, i.e., with cantilever or wire-braced transverse frames and circumferentially spaced longitudinal girders built up from seamless tubing of high-strength aluminum alloys. Resistance to shear is achieved by a diagonal network of steel wiring linking the girder joints. Over this metallic skeleton will be stretched a taut outer cover of aluminized Kevlar fabric. Within the hull, an array of 17 to 18 Tedlar-lined helium cells, each equipped with a set of maneuvering and automatic over pressure valves for the discharge of helium. All living quarters--passenger accommodations, crew quarters, control bridge, etc., will likewise be contained within the hull space so that the only external appendages will be the horizontal and vertical control surfaces, or, as they are called in LTA terminology, the "fins."

Other than the big jump in size over past prototypes, the most important innovation marking a new generation of airships will be its propulsion system. The last Zeppelins were driven by diesel engines mounted in "pods" slung from the sides of the hull. Owing to the rather high specific weights of these power plants, this may have been the only practical arrangement at the time, but it was hardly an efficient one. It was mentioned above that the drag of a well-streamlined hull at large Reynolds numbers consists purely of skin friction. Consequently, an airship drags behind it in flight a wake in the form of an invisible cylinder of air of about half the hull's mean diameter.

On the other hand, each of the propellers driving the airship generates a slipstream, moving in the opposite direction. Both wake and slipstream contain large amounts of wasted kinetic energy which can be chalked up to drag in the one case and low propulsive efficiency in the other. By locating the propellers in the wake, these two sources of lost energy can be made to cancel each other out! Propulsive efficiency will improve dramatically. Fuel consumption will drop sharply--

and payload correspondingly boosted. The huge hull of an airship represents the ideal application of stern propulsion, and along with efficiencies approaching 100% will come other benefits. For example, by incorporating very large diameter, slow-turning counter-rotating propellers the sound level can be reduced by as much as 80-90%. The airship of tomorrow will be the first truly silent ship of the skies - a welcome new-comer to an age plagued by noise pollution. Also, by placing the propellers aft of the fins, the flow of air over the roots of the fins will be improved, thus permitting a reduction in overall fin area.

What type of engines will drive these sixty foot diameter propellers? What energy source will supply them? Here the airship designer will enjoy a flexibility of choice largely denied to his heavier-than-air colleague. For quite obvious reasons, today's jetliner can be powered only by petroleum fueled turbofan engines - miracles of compactness and low specific weight, but noisy and limited by low Brayton cycle thermal efficiency.

A conservative approach to any new airship power plant might be to incorporate an improved diesel engine, with a s.f.c. fully twenty percent lower than that of the best turbofan engine available. This would be equivalent to adding 20,000 pounds to the payload of a 747 jumbo jet on a transatlantic hop--at substantially lower costs per gallon of fuel. But an equally feasible and potentially far more advantageous solution is to make use of the Rankine cycle for airship power plants: a modernized version of the steam turbine energized by liquid hydrogen as the fuel. LH_2 has a heating value almost three times that of hydrocarbon fuels. What's more, it derives from a renewable source--water. Its exhaust products are non-polluting H_2O. And liquid hydrogen is as safe to store and handle as any other liquid fuel. However, since LH_2 boils at 36.7° Rankine, it is a cryogenic fuel and this will entail a fuel system of greater weight and complexity than a conventional one. Furthermore, since the density of LH_2 is low (only 4.43 lb/ft^3) the space which the fuel tanks must occupy is large (a real stumbling block for airplanes, but no great handicap for airships). For airships, at least, the advantages of LH_2 far outweigh its disadvantages and it is uniquely suited for LTA propulsion. In the airship under discussion, the boilers would be located within the lower portion of the hull forward of the fins and the turbines, with reduction gearing, adjacent to the stern-positioned dual rotation propellers. And the condensers--normally the heaviest component of a Rankine system--would be the vast network of tubing comprising the structural girders of the hull itself. Here, rejected heat would be absorbed by the ship's helium to produce additional lift--a real bonus to boot. The effect that all this will have on the payload of a large airship is illustrated in Figure 1.

Once the rigid airship has re-entered the transportation field, its evolution should be rapid. Its uses will undoubtedly be varied, ranging from long-range cargo hauling and scheduled transoceanic passenger service to more exotic uses such as luxury cruise vessels, oceanographic research applications and disaster relief vehicles. Great new hangars will be built, permitting the launching of truly mammoth airships. A hangar two thousand feet in length would suffice for the construction of the hundred million cubic foot behemoth outlined in Table 1. An airship of this size would benefit greatly from a nuclear power plant, as the table indicates. Even the high weight of the reactor shielding would become a minor item in an airship of such proportions, and the added cost of nuclear operation would be easily absorbed by the greater payload. It should be noted that the linear dimensions of this 3,000 ton airship would be only twice those of the 12.5 million cubic foot ship of just one-eighth its volume and lift. With a hull of such large capacity, new structural arrangements other than the traditional space frame would be called for. Transverse frames and longitudinal members would be merged into an integrated trusswork surrounding the helium cells and supporting the outer cover. Shear and radial wiring would be eliminated (except for bulkhead wiring to contain the gas cells.)

Quite obviously, a three-dimensional space frame of such size and complexity will present a major challenge to the designer. Currently, large clear two-dimensional spans are accommodated by the standard flat octet space truss (airport hangars,

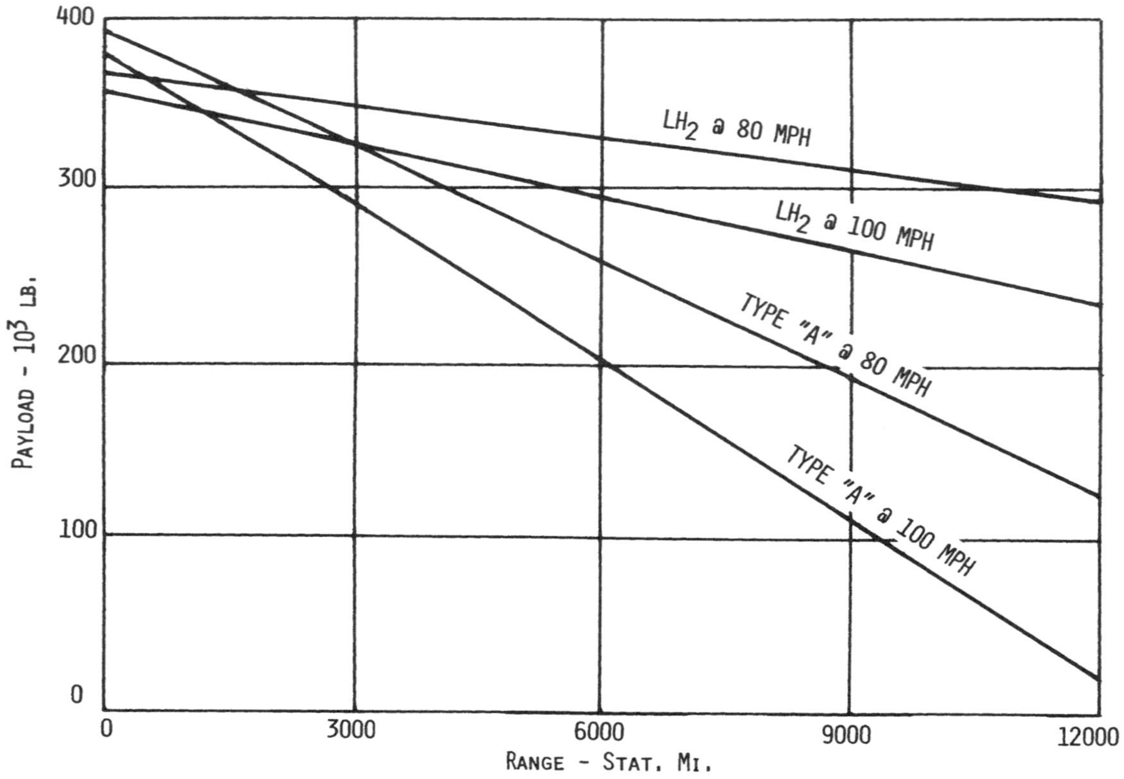

Figure 1. Payload U.S. range for 12.5 million cu. ft. cargo airship using LH_2 or Type "A" jet fuel.

Table 1. Principal characteristics of a 100,000,000 cubic foot rigid airship.

Item	Conventional Power Plant	Nuclear Power Plant
Overall Length, Feet	1,900	1,900
Maximum Diameter, Feet	333	333
Fineness Ration (L/D)	5.7	5.7
Surface Area, Sq. Ft.	1,530,000	1,530,000
Maximum Gas Volume, Cu. Ft.	100,000,000	100,000,000
Volume of Displaced Air, Cu. Ft.	106,000,000	106,000,000
Gross Lift @ 95% Helium Inflation, tons	3,000	3,000
Useful Lift, Tons	2,150	2,000
Payload @ 6,000 Mi. Range, Tons	1,750	1,850
Payload @ 12,000 Mi. Range, Tons	1,500	1,850
Number of Gas Cells	20	20
Number of Engines	8	8
Total Horsepower	32,000	32,000
Maximum Speed, MPH	120	120
Cruising Speed, MPH	108	108
Number of Personnel (Operating)	60	66

sports arenas and the like). Where a doubly curved surface--e.g. the envelope of an airship hull--is concerned, the problem becomes far more difficult. Fortunately, studies in the geometry of folded plate systems by Prof R. Resch of Boston University promise to simplify the procedure of airship hull design and at the same time to offer a feasible method of repetitive element hull construction. Using computer assisted techniques, Resch and Christiansen have defined a series of versatile structural systems which are capable of producing an infinite variety of enclosures (Figure 2). An arrangement of a series of these structures would be eminently suitable for the envelope of a very large airship hull. As well as being both lightweight and strong, the modular pattern would lend itself to economical mass production techniques. Their computer program allows the specification of any surface of revolution. It will proceed to construct a truss tangent to that surface, the depth of which may be specified, and will output a control tape for the description of all the members of the given structure.

Within this gigantic hull would be disposed luxury hotel accommodations for 3,500 passengers; promenade decks longer than the QE2; hangar space for half a dozen shuttle planes or helicopters; cargo holds totalling a million cubic feet or more. Truly, it could be termed the ultimate airship.

A number of structural and aerodynamic advances resulting from enlarged displacement have been described in the preceding paragraphs. A further aerodynamic effect deserves mention here. Airships are noteworthy for their smooth ride -- no case of airsickness in a rigid airship has ever been recorded -- and the larger the airship, the steadier its flight path. This is expressed graphically in Figure 3 which indicates the effect of size increase on the ratio gross lift to surface area in pounds per square foot. At a ratio of 2 lb/sq.ft. of surface area, all motion due to normal atmospheric turbulence is effectively suppressed. This corresponds to a hull length of approximately one thousand feet.

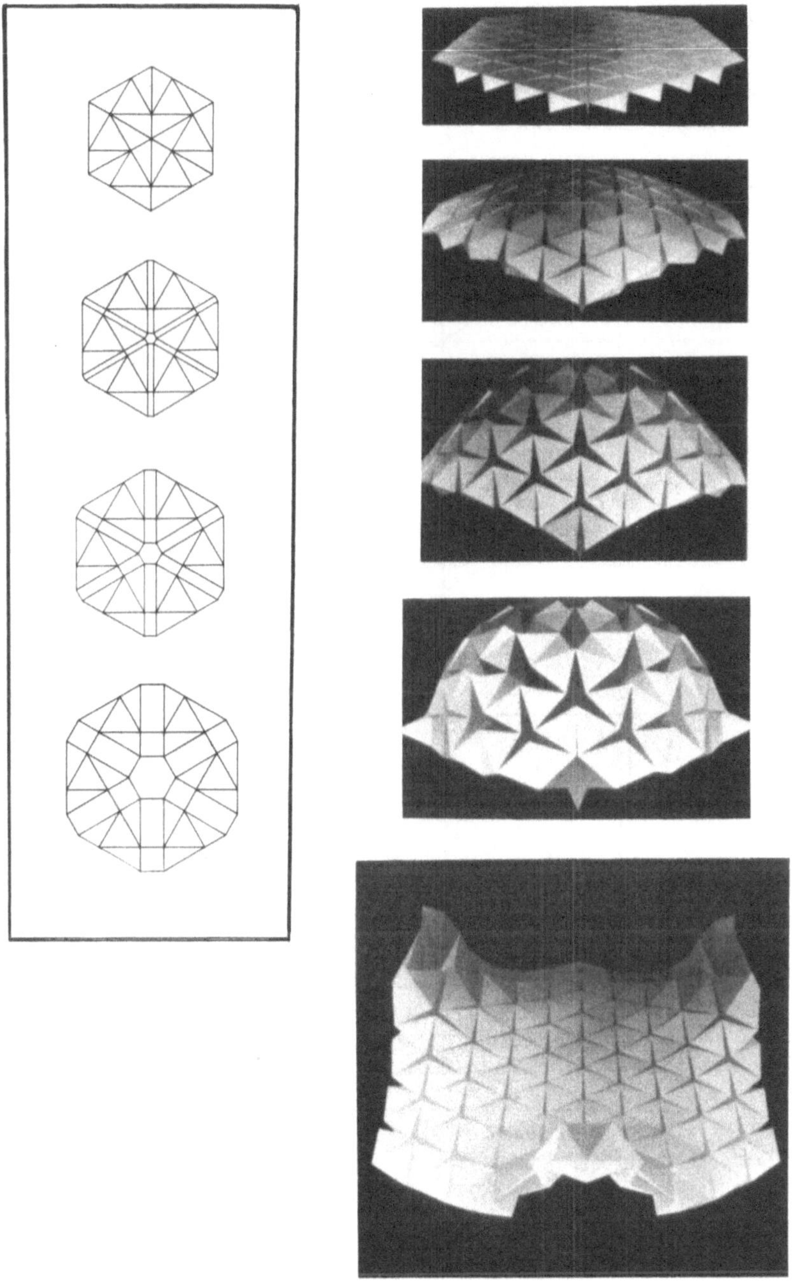

Figure 2. Computer-generated structural systems for various types of enclosures.

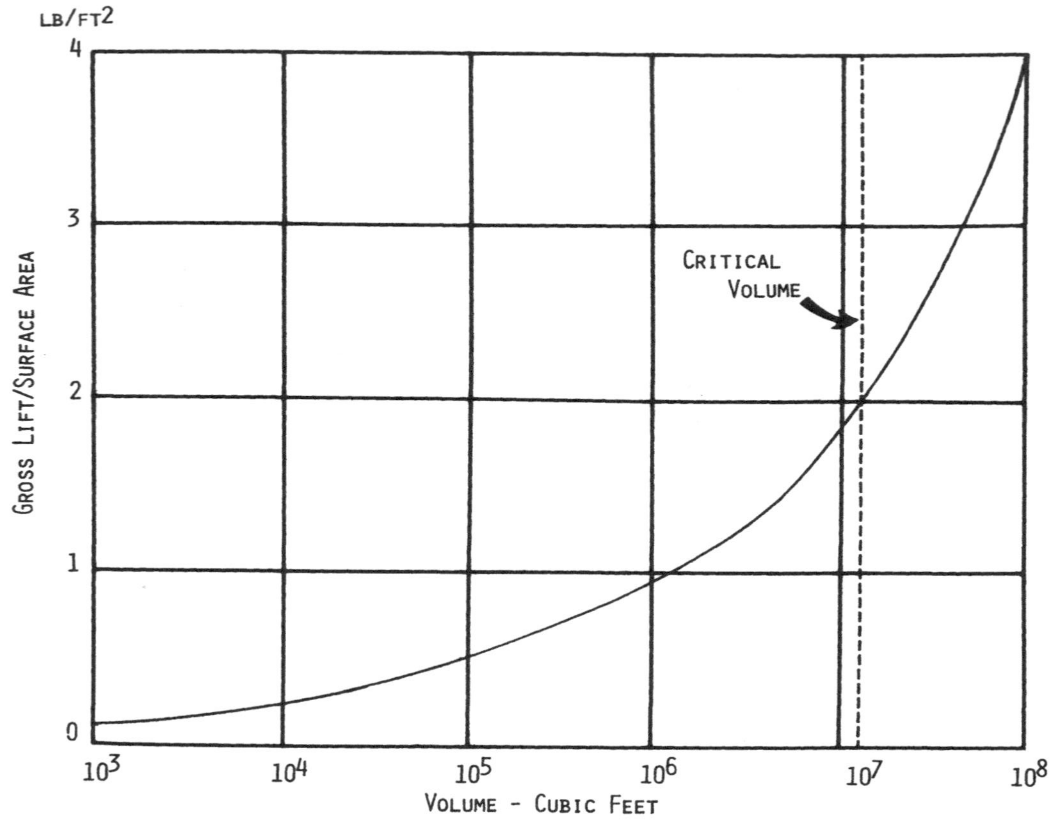

Figure 3. Effect of size increase on ratio of gross lift to surface area.

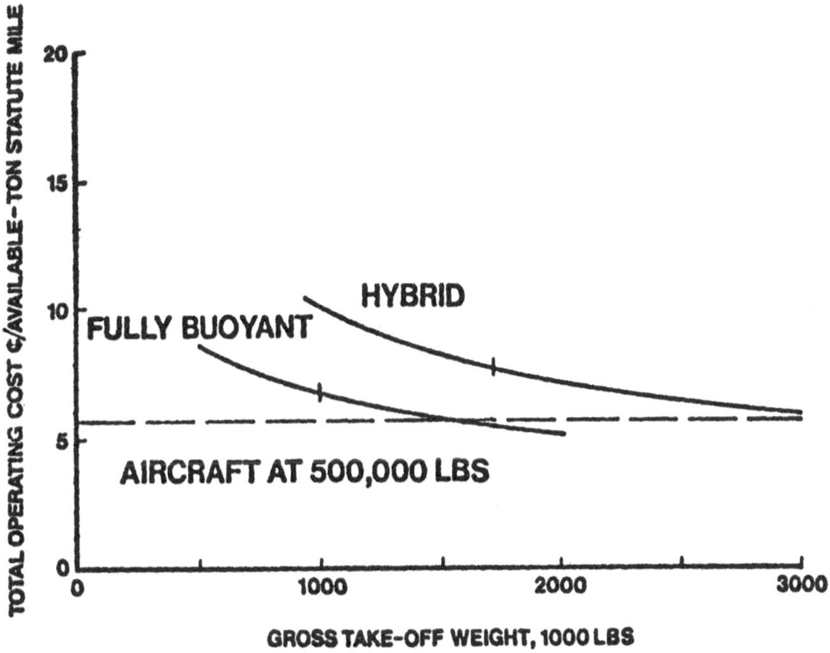

Figure 4. Effect of take-off weight on operating costs for three types of aircraft.

A detailed discussion of airship operating costs is outside the scope of this paper. However, a comprehensive indication of the effect of size on direct operating costs for airships vs. airplanes is shown in Figure 4. Once again, the scale advantage is apparent, mainly because of the decreasing empty weight fraction and improved aerodynamic efficiency with volume growth. In today's competitive society, this alone should be reason enough for the energetic development of new and larger airship types.

Overview

A number of interesting alternatives exist for improved high-speed travel between distant points on the globe. These include electromagnetic levitation trains in vacuum tubes, laser-powered aircraft, hypersonic flight, suborbital rocket-powered vehicles, and orbital transport. J. Peter Vajk stresses that technological, economic, and social acceptance questions still remain to be answered for all of these alternative transport systems. Perhaps the most likely candidate to survive the multi-criteria assessment of these systems is the orbital transport.

J. Peter Vajk

14. Global Transport Systems for the Turn of the Century

The desire to explore the world in which we live appears to be a universal instinct of every living creature. In human consciousness, this urge is experienced as a sometimes painful schizophrenia between the "here and now" and the "there and then." Although it is still possible to find people who have lived their entire lives within just a few kilometers of the house in which they were born, most people on earth today have traveled far more extensively than would be required for physiological survival. In the more affluent nations of the world, we often travel great distances, sometimes in the course of our jobs, sometimes for pleasure as tourists. We casually accept the idea of vacation trips to visit the natural wonders of the world; yet the appreciation we have for natural landscapes, in and of themselves, is a very recent value in Western civilization, having originated with the English Lake poets about the end of the eighteenth century.[1]

Today, extensive networks of highways, railroads, shipping lines, and airlines make travel to far away places far easier than could have been imagined at the beginning of this century. Supplemented by sophisticated networks of worldwide electronic communications, this same transportation system has brought about a progressive increase in the economic, material, political, and cultural interdependence of the nations of the world. Within the last generation or two, the "global village" has become a reality. As our interdependence continues to grow, so will our need for still more efficient transportation systems for long-range travel.

This paper was originally conceived as an assessment of one particular technological extension of the global transportation system, the development and commercial implementation of hypersonic flight. In order to assess the viability of hypersonic flight around the turn of the century, however, it was necessary to consider various alternative extensions of the global transportation system, and it seemed more natural and useful to develop a preliminary comparative assessment of five potential new transport systems which could, in principle, be implemented by the turn of the century.

The Need for Global Transportation

What purposes would be served by improving the global transportation system beyond the present technology represented by such aircraft as the Boeing 747 and the Concorde SST? Why should anyone want to travel any faster than these sleek mechanical birds can fly? Is hypersonic flight the next logical step in the evolution of the global transportation system?

Let's consider the uses of the global airline system as it stands today. Tourism, of course, is an important function of this system, providing a great deal of personal satisfaction, pleasure, and enrichment to many. Air travel has enhanced our knowledge and appreciation of the biosphere and of the diversity of human culture and experience. For most inhabitants of the New World, it has made it possible to reestablish our appreciation for our own ancestral and cultural roots. The recent breakthroughs in trans-Atlantic fares triggered by Sir Freddy Laker have already expanded the personal experiences and educations of countless students who can now afford to visit Europe. The human value of alleviating the pain of wanderlust is not to be scorned, either.

But on a routine, round-the-year basis, the backbone of the airline industry's revenues comes from the business and government sectors. Air freight shipment of high priority cargoes, air mail service, and air travel by business executives, sales representatives, and government officials are of critical importance to the day-to-day operation of the interdependent global economy. In the last few years, this has become especially visible in the form of thousands of American managers and engineers who are helping some of the major oil-producing nations of the world (Saudi Arabia springs to mind here) to develop their resources and to bootstrap themselves into the modern age.

The business executive who can now return from Europe aboard a Concorde the same afternoon of his (or her) departure from New York, carrying signed contracts worth several million dollars, has become a routine element of international trade. But the value of such commercial uses is not exclusive to the industrialized, affluent nations. Such travel is enormously valuable to business and industry in the Third World as well, since it has given them access to trade fairs and global markets which were scarcely aware of the existence of many companies as recently as ten or twenty years ago. In the absence of personal business contacts, Third World companies could scarcely participate (let alone compete) in the international marketplace.

"Shuttle diplomacy" is a phrase coined in the 1970's to describe the effective use of rapid air transportation between the political centers of the world. While shuttle diplomacy was most dramatic in the Middle East where air travel times are as little as an hour or two, the current transition in U.S.--China relations (whatever one's opinion of the political wisdom of that transition) could not have been effected nearly as rapidly in the absence of the global air transportation system.

The global transportation network has also facilitated major international development programs aimed at alleviation of human misery in the most impoverished parts of the world. Disaster relief, famine relief, and evacuation of refugees have occasionally relied on massive airlifts. Air transportation on a more routine basis has brought expertise, equipment, and managerial skills into underdeveloped areas, while permitting students, scholars, business executives, and government officials from Third World countries to study, consult, and learn abroad. Although the rapidly expanding electronic communications networks have made immeasurable contributions to this process as well, the electronic media are not--and never will

Table 1. Some major cities in the global transportation network.

Berlin	Lisbon	Peking
Bombay	London	Rio de Janeiro
Buenos Aires	Los Angeles	Rome
Cairo	Manila	San Francisco
Calcutta	Mexico City	Santiago
Capetown	Miami	Seattle
Caracas	Montreal	Shanghai
Chicago	Moscow	Singapore
Colon	Natal	Stockholm
Dakar	New Orleans	Sydney
Hong Kong	New York	Tokyo
Honolulu	Novosibirsk	Warsaw
Istanbul	Paris	Washington
Khabarovsk		

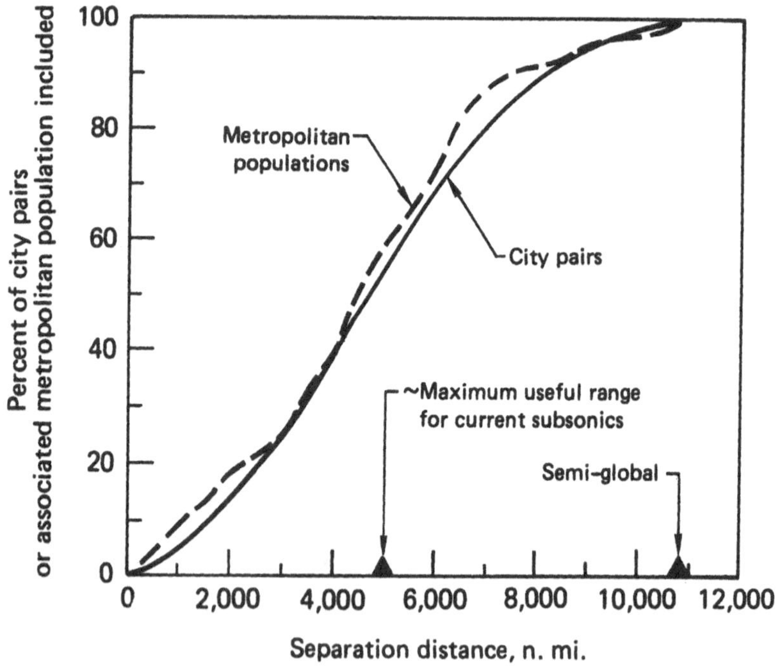

Figure 1. Sample distribution of inter-city distances in a globally interdependent economic system. Adapted from Robert Salkeld, "Global Rocket Transports: Changing Perspectives," in The Second Fifteen Years in Space, Vol. 31, Science and Technology, American Astronautical Society, San Diego, 1973.

be--a complete substitute for personal face-to-face discussions and for on-site experience.

While present day jet aircraft are far more reliable than their piston-engine predecessors, mechanical failures and breakdowns do occur on occasion. Should this happen to an airplane in Boston or London or Frankfurt, repair facilities, personnel, and replacement parts are readily available on the field, and the flight may be delayed an hour or so. But if a jumbo jet has a mechanical failure in Montevideo or Nairobi or Karachi, chances are high that the replacement parts must be flown in from the United States or Europe, in some cases with tools and mechanics as well. The delay before that aircraft can return to service may then be 24 to 72 hours, because of the lengthy travel time for the parts, tools, and repair personnel. Where alternate flights are infrequent or have few empty seats, most stranded passengers in this situation would also be delayed up to three days behind their schedules, and the airline's entire timetable may be severely disrupted. Economic losses can total far more than the direct inconvenience to a few hundred passengers.

All of the present uses of the airline transportation network around the globe would clearly benefit from improvements in its effectiveness and capabilities, especially as longer range trips become more and more common, up to trips half-way around the globe.

The maximum practical non-stop range for subsonic jet transport aircraft in use today is about 5,000 nautical miles (5,750 statute miles or 9,200 kilometers). As shown in Figure 1, nearly half of the 40 major airline hub cities listed in Table 1 are separated from each other by more than this distance. The 40 cities listed are not the largest in population, but were included on the basis of their present character as regional hubs for airline transportation, size of population, and political, economic, and cultural importance. The distances between cities for each of the 780 different pairs of cities were computed as great circle distances; because of lack of navigational facilities or because of political necessities of flying around certain territories, actual air route distances are in all cases somewhat greater than these great circle distances.[2]

Figure 2 shows the travel times associated with long-distance travel by present day subsonic jets and by the Concorde supersonic transport. Depending on the traveler's personal decisions (or on available flight schedules), traveling half-way around the world today by subsonic jet provides a choice of two different ordeals by endurance. Either one flies straight through, with one or more refueling stops or flight connections, for a twenty-four hour or longer siege; or one breaks the trip with an overnight stop, which almost invariably involves clearing customs and immigration, ground transportation to the hotel district, partial unpacking of luggage, repacking after a poor night's sleep, transportation back to the airport, all of this compounded by the possibilities of travel time of 35 or more hours. The situation is somewhat improved for supersonic transport, with the possibility of getting half-way around the globe in less than ten hours if flight connection times or refueling stops are short and uneventful. Breaking the trip with an overnight stop results in a total travel time of a bit more than twenty hours. Note that the times shown in Figure 2 are optimal--they assume great circle routes can be flown; that headwinds are moderate; that connection times or refueling times are ideal; and that flights are available just when the traveler wishes to go.

Figure 2 also shows travel times for a Mach 8 hypersonic transport and for a global orbital rocket, but I want to defer discussion of these possibilities until later. Note, however, that the hypersonic transport would allow semi-global distances to be covered in a total travel time of as little as five hours, a time which is commonplace today for transcontinental flights in North America today by subsonic jets. Travelers on such flights do experience a significant level of fatigue and stress, but the experience is seldom perceived as an ordeal of endurance.

Global Transport Alternatives

What are the possible options for extending the global transportation network toward more

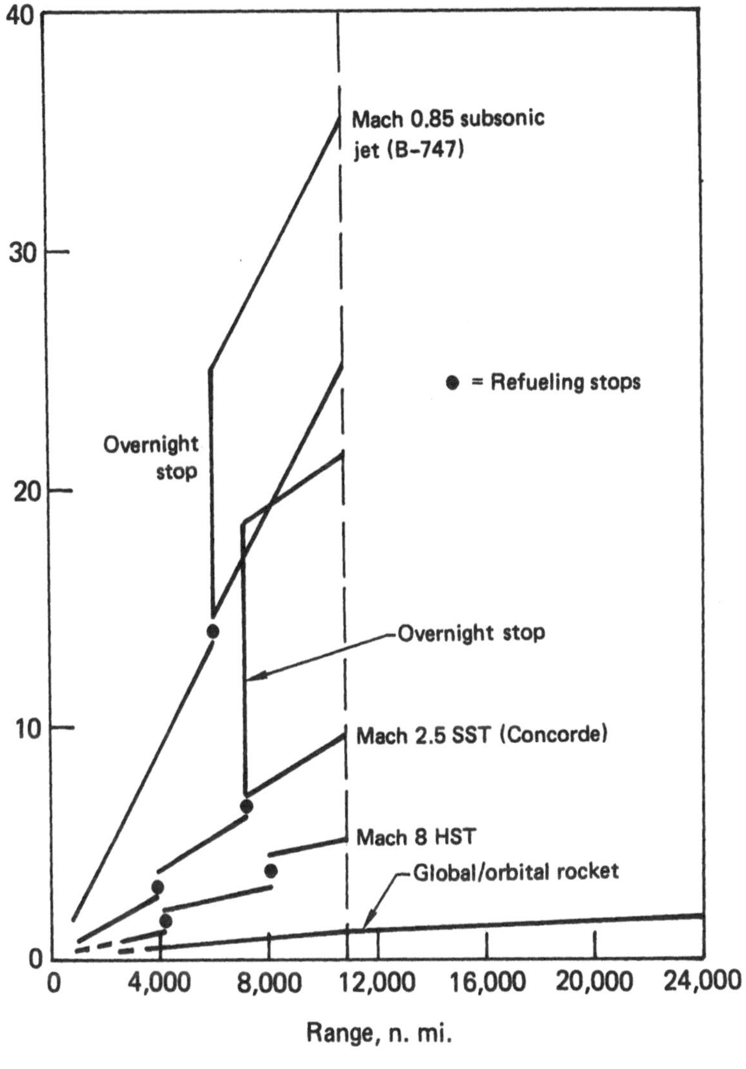

Figure 2. Travel time as a function of distance by current and projected flight technologies. Adapted from Robert Salkeld, "Global Rocket Transports: Changing Perspectives," in The Second Fifteen Years in Space, Vol. 31, Science and Technology, American Astronautical Society, San Diego, 1973.

efficient and rapid transportation between two cities more than 5,000 nautical miles apart? Obviously, transportation paths between two cities on the surface of the earth may lie beneath the surface of the planet, on the surface, in the atmosphere, or through the atmosphere into space and back. Before discussing new options, I want to indicate briefly some of the reasons why surface transportation cannot be expected to contribute to a major extension of the global transportation system.

Land transportation requires smooth roadways if it is to be fast and safe. On natural surfaces which are reasonably smooth (such as snow and sand) sliding vehicles have been used since prehistoric times, but these are very inefficient because of friction losses. Tracked vehicles have proven very useful on uneven surfaces, but their mechanical complexity restricts their speed and reliability. The biological solution of articulated legs has been highly successful in the animal kingdom for hundreds of millions of years, but only because biological computers are sufficiently compact to perform the complex calculations needed to control the motion and placement of the legs on irregular surfaces and to interpret the kinesthetic feedback required for even stress distribution among the legs. Aside from the mechanical complexities of artificial articulated legs, our computer technology will be inadequate to this task for many years to come, even for very modest speeds of travel.

The historical solution for efficient land transportation has thus been wheeled vehicles on specially prepared roads or rails. The speed of transportation has been bought dearly, however, since it requires a major investment of resources to build the network of roadways and to maintain it. In the United States, we have spent approximately $175 billion (in 1978 dollars) to construct the Interstate Highway System over the last three decades, and much greater sums for state, county, and municipal highways, roads, and streets. Wheels on rails can be more efficient still, especially if all grades are moderate, permitting higher speeds in principle.

Could newer technologies permit really high speed long range rail transportation on the surface? Any such scheme must come to grips with four basic problems of wheels on rails: lateral stability, aerodynamic drag, intrinsic strength of the wheels, and security of the right-of-way. As train speeds become higher, irregularities in the track and motions of passengers or cargoes inside result in oscillatory motions of the railroad cars. If these motions are not to result in excessive lateral forces between the wheels and the rails or in unstable oscillations, the passenger compartment must be more tightly coupled to the rails and the tracks must be more precisely aligned, despite the greater lateral forces at higher speeds. This problem has restricted the actual operating speeds of the Japanese railroads to substantially less than their original design speed. The upper limit for the foreseeable future seems to be less than 200 miles per hour for advanced rail systems.

Even at speeds of 200 mph, however, trains are at a disadvantage compared to airplanes because of aerodynamic drag, since airplanes can cruise at high altitudes where lower air density reduces drag far below that felt by streamlined trains at sea level. At still higher speeds, the centrifugal stresses on wheels, in combination with the enormous lateral forces sustained by the wheels in holding the train on the tracks, become dangerously close to the tensile strength of high grade steels.

These difficulties with wheels can be avoided by air cushion vehicles or by magnetic levitation. At high speeds, however, both of these technologies must still face the problem of aerodynamic drag. The necessity in air cushion vehicles for large air intake scoops and a wide base is sharply at odds with the needs of streamlining the entire vehicle. At high speeds, maintaining the stability of an air cushion vehicle to prevent accidental contact between the vehicle's structure and the roadway poses a very severe problem.

Any surface roadway or railway intended for high speeds has a perennial problem in maintaining the right-of-way secure against intrusion by pedestrians, errant animals, or debris. Fencing is (at best) a partial solution, far from failsafe.

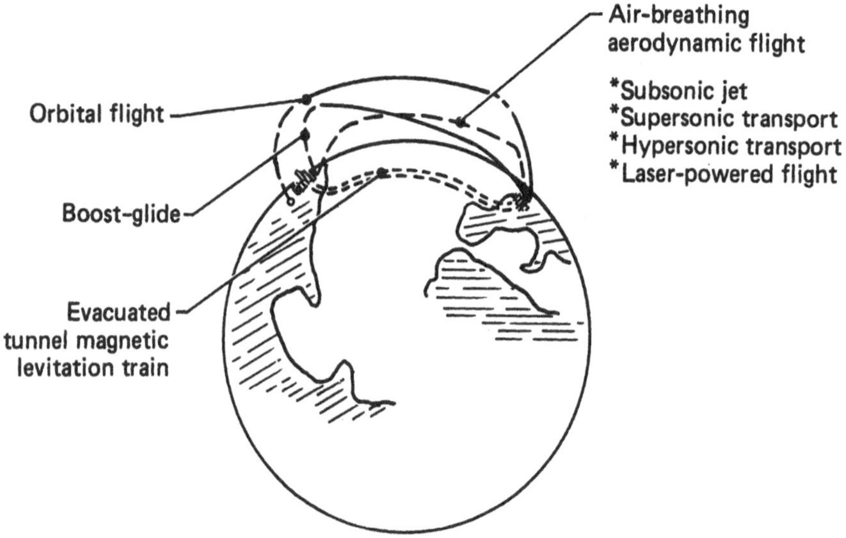

Figure 3. Global transportation alternatives for the turn of the century.

Pieces of metal thrown over a fence or dropped on the right-of-way from an overpass, or large tree branches blown onto the right-of-way by storms can derail trains or rupture tires, producing devastating accidents.

Transportation by waterways and by sea has been used extensively all through history. Its principal use today is for the transportation of bulk cargoes of relatively low value per unit mass or per unit volume (grains, lumber, textiles, fossil fuels). The principal advantage of aquatic and marine transportation is the fact that open water provides a level surface: little or no effort must be expended in carrying the cargo up and down the terrain. For high speeds, however, both surface and submersible vessels are severely limited by the enormous energy required to overcome hydrodynamic drag: the engines must provide the energy to displace enormous masses of water in front of the vessel to permit its passage. This difficulty can be minimized by hydrofoils, but present day experimental hydrofoils operated by the U.S. Navy can achieve only about 100 knots. Little prospect for major improvements is in sight, so hydrofoils are unlikely to present significant competition even to subsonic jets in the long-range transportation field. Ships using the air cushion effect would eliminate the hydrodynamic drag completely, but the problems of aerodynamic drag at sea level would limit marine transportation just as much as land transportation at high speeds.

It thus appears that high speed long range transportation on the surface of the earth has little future. What options remain? Besides existing subsonic and supersonic aircraft and some improvements on the existing system, five new systems could conceivably be implemented on a significant (if not fully global) scale by the turn of the century. These five technologies, shown in Figure 3, are (1) evacuated tunnel magnetic levitation trains running underground and underwater; (2) hypersonic aerodynamic flight with air-breathing engines; (3) air-breathing aerodynamic flight using laser beams for power; (4) rocket-powered flight in the atmosphere, operating in a boost-and-glide mode; and (5) rocket-powered flight into space and return. Each of these will be discussed below.

A Global Subway System

The difficulties of surface land vehicles discussed above can be overcome by going underground to develop a novel transportation system. The problem of security for the right-of-way and the adversities of foul weather and large temperature changes every day are eliminated by digging tunnels through solid rock a few hundred feet to a few thousand feet below the surface. Since the entire path can be completely enclosed, the tunnel can be evacuated to 10^{-3} of sea level pressure and density allowing sealed vehicles to reach very high speeds with negligible aerodynamic drag. (This pressure and density are equivalent to those in the earth's atmosphere at 170,000 feet above sea level.)

The problems of wheels on rails can be eliminated by substituting electromagnetic levitation and acceleration for more conventional suspension and propulsion methods. Each vehicle would be equipped with several superconducting coils carrying large currents. The currents, once initiated, would be maintained by small cryogenic systems. The cars of a train would then be accelerated and decelerated by magnetic fields induced in drive coils mounted in the tunnel walls. The timing and shape of the current pulse through each drive coil would be controlled by microcomputers which obtain information on the position and velocity of each car from optical and/or electronic sensors along the tunnel. Kinetic energy pumped into the train as it is accelerated can be recovered with high efficiency when the train is decelerated by the drive coils, in a kind of regenerative braking.

A fairly detailed conceptual design for such a system was presented in the 1978 Macro-Engineering Symposium.[3] The state-of-the-art in electromagnetic levitation and acceleration is perhaps most advanced in the work of Gerard K. O'Neill, Henry Kolm, and their associates on an electromagnetic catapult for launching raw materials from the surface of the moon for use in space manufacturing.[4]

What sorts of speeds could such a global subway system achieve? This is principally a question of how high an acceleration passengers can comfortably accept and of how much power can be delivered to the drive coils in the tunnel walls to accelerate the trains. In practical engineering terms, sufficient power to accelerate the trains at several g's can be provided, with higher accelerations implying higher costs for hardware. The dominant factor in the total cost of the system, however, is the cost of digging the tunnels. Thus, the question of passenger acceptance of acceleration is the real criterion. If one-third g is acceptable, then such a train system could provide a total travel time of 27.3 minutes for the 2700 statute mile trip from downtown New York to downtown Los Angeles via Dallas, reaching a peak speed at the midpoint of the journey of about 6000 mph.

A similar acceleration schedule on a tunnel following a great circle route halfway around the globe would provide semi-global travel times of 82.5 minutes, reaching a peak speed of 18,100 mph at the midpoint of the journey before commencing deceleration! At this speed, passengers would experience slightly more than one-half g centrifugal acceleration outward from the center of the earth, so their weight would be slightly less than half of normal. It has often been asserted that the fastest we could ever hope to travel on earth is orbital speed, 25,000 mph. But it is clear that tunnel trains could actually go faster than this: at 35,380 mph, passengers and cargoes would be pushed toward the top of the train at exactly one g, so that if the entire train turns upside down while accelerating through orbital speed, it could continue to accelerate to this higher speed. At that point, passengers would experience normal weight, albeit they would be seated or standing upside down! If we consider again a trip exactly halfway around the globe in such a tunnel, this peak speed could be reached with a constant acceleration of 1.27 g for 27 minutes followed by deceleration for the same time.

Because of centrifugal forces, the tunnels would have to follow great circle routes, at least in their high speed segments, and follow very precisely a smooth arc about the center of the earth. Grade changes would have to be very gradual to avoid the sensation of riding a very severe roller-coaster.

If we constrained such a system to lower accelerations, lower peak speeds, and shorter segments between stations enroute, it would still be possible to match the travel times shown in Figure 2 for hypersonic transports for long distances, with the major advantage that the stations for a global subway could in almost every case be located in the heart of the downtown regions of cities on the subway network.

Such a system is certainly feasible from an engineering and technological point of view by the turn of the century. But the economic feasibility of the system is an altogether different question. Such a system for the United States, with major lines connecting New York to Dallas and to Los Angeles, major lines connecting Chicago to New York and to Dallas, and feeder lines reaching out to San Francisco, New Orleans, Kansas City, Buffalo, Montreal, and Portland, Me., among other cities, has been estimated to cost about $250 billion.

The capacity of such a system for the United States would be five to ten times greater than the air traffic projected for the 1990's. To make the system economically viable, it would then appear necessary to integrate the subway system with other transportation and transmission systems which can advantageously use underground tunnels. Conventional rail freight operating at high speeds with minimal grades and no grade crossings is a distinct possibility, especially if used for transportation of Western coal. The underground Fast Railway System, directly connected with conventional railroads above ground, would offer a better alternative for a coal-based electric utility industry than 100-car long unit trains running at ten to fifteen minute intervals on the surface.

Other functions which could be integrated with the transcontinental segments of a global subway system include oil and gas pipelines, superconducting electrical power transmission lines, and long distance laser communications lines. With several such systems sharing the costs of the initial tunneling, fares of $50 per passenger or less for coast-to-coast travel would be possible. But the legal problems posed by the full tangle of anti-trust legislation and of conflicting bureaucratic jurisdictions and regulations involved in attempting to integrate such a diversity of systems under a single project might easily consume decades to resolve. A further legal issue of a new kind is the question of compensating property owners for the use of the subterranean right-of-way, since property deeds, in common law, have usually granted absolute rights to a land owner to everything included in a cone down to the center of the earth. Subways in major cities have generally been built under publicly owned land, so little precedent is available for determining the value of these rights.

Atmospheric Flight

In comparison with landgoing vehicles, flying machines have always enjoyed infinite flexibility in routing, since prepared surfaces are required only at the departure and destination points. Seagoing vessels, of course, enjoy this advantage too, but the density of air being about a thousand times less than that of water, aerodynamic drag is much less than hydrodynamic drag, so that much higher speeds are practical with reasonable engine power and energy consumption rates.

Figure 4 shows the flight corridor accessible to various types of flying machines.[5] At any given speed, the aircraft is constrained to a certain range of flight altitudes, either because the air is too thin to provide sufficient lift at higher altitudes, or because the air is so thick at lower altitudes that friction between the air and the aircraft skin produces excessively high temperatures. Both limits are subject to some expansion, as some room for improvement in the design of wings and lifting bodies still remains, and as structural materials which can withstand still higher temperatures become available.

The propulsion system used also depends on airspeed and altitude of flight. At low speeds in dense air, piston engines and propellors are efficient and effective, although less reliable than jet engines. As the speed of flight increases, the tips of the propellor blades approach the speed of sound, and the shock waves propagating backward have adverse effects on the stability, controllability, and comfort of flight,

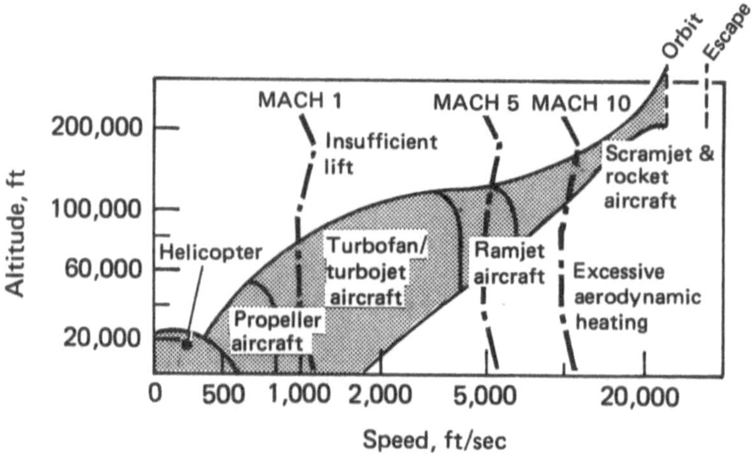

Figure 4. Flight corridor as a function of airspeed. Adapted from William T. Hamilton, "Projected Aircraft Systems Development," in The Future United States Space Program, Vol. 38, Advances in the Astronautical Sciences, American Astronautical Society, San Diego, 1979.

while the overall efficiency of the propellor itself is seriously degraded. At higher altitudes, piston engines become less and less powerful because each stroke contains a smaller total mass of fuel and air and thus releases less energy per combustion cycle, unless the intake air is compressed by a turbocharger. At still higher altitudes, however, the temperature of the compressed air rises too high for engine components to tolerate, since the temperature depends on the compression ratio, not on the final density. If the intake air is less and less dense, the compression ratio must rise higher and higher to produce the final density needed to sustain power in the combustion cylinders.

Turbine engines perform very well over a wide range of altitudes and a significant speed range. Compression of the intake air is achieved partially by narrowing of the air flow laterally from the engine intake through the compressor into the combustion chambers, the very high-temperature mixture of combustion products expands rapidly through the exhaust nozzle to attain speeds much higher than that of the airstream past the aircraft as a whole to provide thrust. Up to airspeeds of Mach 3 or 4, then, the flow of the air/fuel mixture through the combustion chambers remains subsonic, even if the inlet and outlet speeds are supersonic. At higher speeds, however, it is impossible to keep the flow subsonic without going to very high compression ratios and thus to unacceptably high engine temperatures. In addition, shock waves from the supersonic flow impinging on the forward portions of the compressor begin to impair the efficiency (and even the mechanical integrity) of the later compressor stages.

At higher speeds, ramjets take over as the most effective engine. Having no moving parts, ramjets are far less vulnerable to internal shock waves. By narrowing down the channel for air flow through the engine, the ramjet compresses the intake air laterally and also longitudinally by slowing the flow of air through the engine. Thus, although the intake airflow may be supersonic, the flow through the ignition and combustion regions remains subsonic, again with a supersonic exhaust speed. Ignition results solely from the high temperature of the compressed air and fuel mixture.

But as the airspeed increases still further, the flow through the combustion region itself becomes supersonic, and conventional ramjets become prone to flameout. Supersonic-combustion-ramjets (usually called scramjets) must take over the task at airspeeds above Mach 6 or 7, the threshold for hypersonic flight. Note that both ramjets and scramjets cannot operate at all at speeds below Mach 2 or 3: at such low speeds, the intake air is insufficiently compressed to ignite the air/fuel mixture. Thus, aircraft designed for very high speeds must be equipped with two sets of engines: conventional turbine engines for takeoff, climb and acceleration to sufficient speed for the ramjets or scramjets to ignite. The conventional engines can then be shut down and their intakes closed over to reduce drag.

An alternative engine for high altitude, high speed flight is the rocket engine, which is intrinsically far simpler than any of those discussed above. The flow of fuel, oxidizer, mixture, and combustion products remains subsonic throughout the engine, until the exhaust gases pass through the expansion nozzle to produce thrust. The *a priori* liability of rocket engines in contrast to the air-breathing engines we have discussed so far is the necessity of carrying along a supply of oxidizer as well as of fuel.

Laser-Powered Flight

In view of the rising costs of fossil fuels, considerable attention has been devoted to improving the fuel efficiency of airplanes. Since the only purpose of the fuel burned in turbine engines is to heat the intake air to sufficiently high temperature to produce thrust when it re-expands through the exhaust nozzle, it is perfectly logical to eliminate the fuel altogether and to heat the compressed air by means of a remote energy source which need not be carried aboard the airplane. The idea of powering airplanes by high power lasers in space has thus been advocated by a number of studies[6] and at least one patent has been issued for this concept.[7]

Figure 5 illustrates one version of this transportation system.[8] A satellite located in a sun-synchronous orbit at 1500 km altitude at a 97°

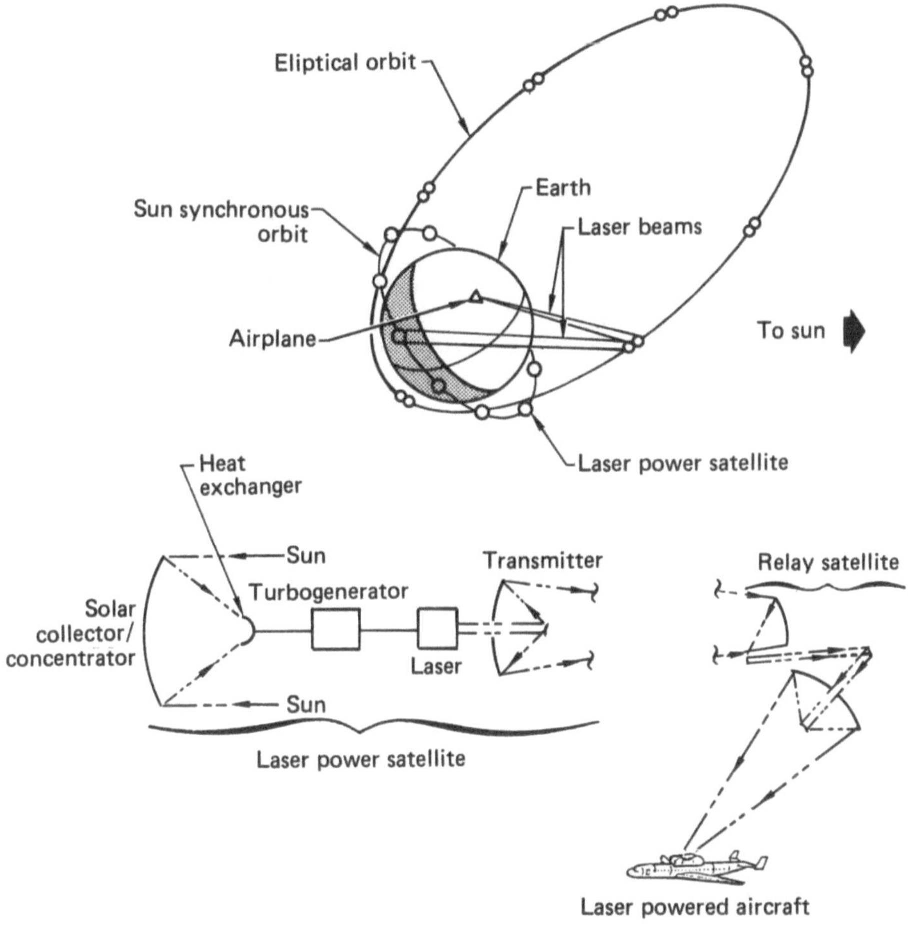

Figure 5. Schematic illustration of laser-powered flight system. Adapted from K. Sun, A. Hertzberg, and Wayne S. Jones, "Laser Aircraft Propulsion" in The Future United States Space Program, Vol. 38, Advances in the Astronautical Sciences, American Astronautical Society, San Diego, 1979.

inclination to the equator captures solar energy, converts it into electrical energy to power a bank of high power lasers, and directs the laser beam to a relay satellite in a 4-hour elliptical orbit at a 63.4° inclination with a 500 km perigee and a 12,300 km apogee. In such an orbit, the relay satellite is above the northern hemisphere for three-fourths of each orbit, and is thus in a position to power airplanes in the part of the world where the vast majority of the air traffic presently flies.

Each airplane carries a 15-meter diameter receiver on its upper surface which focuses and directs the 10-meter diameter beam of laser radiation into a blackbody absorber cavity. The thermal energy is then transferred through a heat exchanger to the air flowing through the turbine engines, with the heat exchanger located between the compressor stage and the conventional combustor. In normal operation, the airplane would take off and climb to cruise altitude with kerosene fuel and then signal the satellite system, requesting assignment of a laser beam. Within a few seconds, a relay satellite could acquire and track the airplane and lock a laser beam onto the receiver. The kerosene flow to the engines would then be shut off allowing the airplane to cruise at stratospheric altitudes indefinitely. (Such a system would clearly have major importance for military applications, as a forward reconnaissance and observation platform, as an airborne command post, or as a deterrent strategic bomber.) When the airplane approached its destination, the flow of kerosene would be resumed and the engines reignited, the satellite system advised that the laser beam was free for reassignment, and the airplane would begin its descent. Should the laser beam be lost inadvertently due to a system failure of some kind, the fuel onboard would allow the airplane to fly 575 miles before landing, assuring a comfortable safety margin for most operations.

The fuel savings, obviously, would be substantial in comparison with conventional jet transports. Assuming a fleet of 300 laser-powered airplanes used in transcontinental and intercontinental traffic, the savings in kerosene in one year would repay the energy investment in deploying the 300 laser power satellites and 400 relay satellites needed for this system. (Each laser power satellite is estimated to have a mass of about 740 tons, while the relay satellites would be 85 tons each.) Total cost of such a system is estimated (in 1978 dollars) to be about $52 billion for the space segment and $1.6 billion for the fleet of 300 airplanes, which is slightly more than one-tenth of the size of the present domestic airline jet fleet in the United States.

Assuming a useful life of thirty years for the satellites and for the airplanes; replacement of 10% of each satellite's mass in the course of maintenance over its life; 6 to 8% annual interest burden; and equal annual payments over the life of the system, the system would be economically competitive for intercontinental flights (more than 7500 kilometers or 4680 miles) if kerosene prices rise above $1.40 per gallon, and for transcontinental flights (more than 5500 km or 3345 miles) if kerosene goes above $2.00 per gallon.

For really long range trips (up to semi-global range), the main time saving associated with laser-powered flight at subsonic speeds would be the elimination of enroute refueling stops. Sleeper service might then become very attractive on long flights. (Japan Air Lines has recently begun to offer sleeper service on some of its trans-Pacific flights).

Environmentally, laser-powered aircraft appear very attractive, since the airplanes leave no exhaust products in the stratosphere during cruise. With failsafe feedback controls on the aircraft tracking and beam pointing systems, the laser beam could be shut off within less than 100 milliseconds after wandering off the receiver on the airplane, so little or no laser radiation should ever reach the ground. In normal operation, the laser beam would sweep across the earth at the cruise speed of the airplane, so that inadvertent exposure to the beam would be very brief, and no unwitting target on the ground should ever be exposed to more than 1 Joule/cm^2 of infrared laser light.

The principal obstacle to such a system is the large initial capital investment required to deploy the space segments. The total amount of financing required is probably beyond the capability of the entire domestic airline industry, should anti-trust rules even permit them to collaborate on such a venture. Either an entirely new corporation would have to be formed to undertake the project--most likely with government guarantees or subsidies--or the federal government itself would have to carry out the program. Needless to say, the political obstacles would be substantial. Such a system would certainly be much more viable as a spinoff or subsystem of a Solar Power Satellite program based on laser transmission of power to the earth for use by the utility industry as a primary source of baseload generating capacity.

Hypersonic Flight

We saw in Figure 2 that the Concorde SST, with a cruising speed of about Mach 2.5, can provide a total flight time of about ten hours for semi-global range. To cut this time in half, it is necessary to aim at a cruising speed substantially more than twice as high because of the time spent on the ground during refueling stops and the reduced airspeeds during climb and descent. A hypersonic aircraft capable of cruising at about Mach 8, having slightly longer range than the Concorde, would reduce the time for a trip halfway around the globe to about five hours, as shown in Figure 2. While the Concorde cruises at altitudes of 40,000 to 60,000 feet above sea level, a hypersonic transport (HST) would cruise above 90,000 feet, as indicated in Figure 4.

The flight aerodynamics for a hypersonic transport are basically reasonably well understood, due in large measure to the experimental rocket airplane program, beginning with the X-1 up through the X-15. This flight regime has also been studied extensively as part of the development program for the Space Shuttle which reenters the perceptible atmosphere at speeds in excess of Mach 20, decelerating as it descends generally within the flight corridor shown in Figure 4, although during the early stages of the reentry it dips below the corridor for a brief time, requiring thick ceramic tiles for thermal protection on the undersurface which is exposed to the greatest heating.

Generally speaking, an aircraft designer will gear the design toward the lower side of the flight corridor shown in Figure 4. Although the viscous drag which produces heating is greater at the lower altitudes, the induced drag associated with the aerodynamic lift is generally less since the total surface required to generate enough lift to support a given gross weight in flight is smaller at higher air densities. If the lifting surface is smaller, the induced drag (which is generally much larger than the viscous drag) will also be smaller, and thus, the airplane will require less fuel for a given flight range. The cost of fuel economy is greater heating on the outside of the airplane, and this poses a significant but not overwhelming problem for HST design. Present thinking suggests the use of liquid hydrogen (cryogenic) as the principal fuel; because of the low temperature of the liquid hydrogen, the fuel lines from the tanks to the engines can run past the hottest parts of the aircraft skin to provide active cooling. (The fuel would also be used in the same way to cool the engine walls.) Passive techniques for keeping the cabin temperature reasonable would also be used, including multi-layered composite wall structures such as honeycombs.

The development of an operational scramjet is the principal engineering problem for hypersonic transportation. During the period 1965 to 1975, NASA conducted a Hypersonic Research Engine (HRE) Project to demonstrate the feasibility of achieving high internal performance with flight-scale scramjets. A podded, axially symmetric engine was developed and tested in wind tunnels at up to Mach 7 conditions. Due to the large nozzle expansion required at high speeds and to shock losses associated with engine struts, it was recognized that the external drag on such an engine would be very high. Thus, if an engine built in such a configuration were to be installed on a hypersonic airplane, overall performance (net thrust) would be poor. NASA had planned to test the HRE in actual flight, using the X-15 as a testbed, but the X-15 program was terminated before the HRE could be flown.

The HRE experience showed the necessity of considering engine design in conjunction with the airframe design. The presently favored concept is

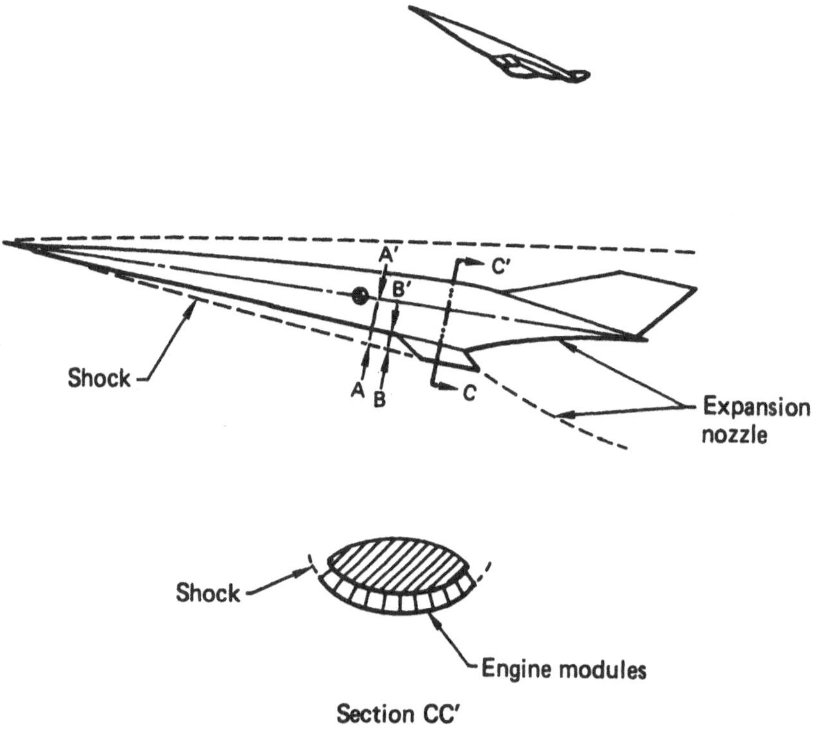

Figure 6. Airframe-integrated scramjet concepts. Reprinted with permission from the American Institute of Aeronautics and Astronautics, Astronautics & Aeronautics, February 1978, pp. 38-48.

the "airframe-integrated scramjet," whose basic ideas are shown in Figure 6.[9] In this airplane design, the shock wave from the nose of the plane just touches the bottom of the engine inlets when the airplane is at its design cruise speed. Most of the airflow which would have passed through segment AA' has been compressed vertically into BB'. For a typical airplane dimension, the vertical height of the scramjet inlet BB' might be 1.5 to 2 feet, midline might be 9 to 12 feet, resulting in a six-fold compression of the air entering the engine. If modular engines fill up the underbody sections of an annulus projecting out from the aft portion of the fuselage to the shock surface, the engines can capture virtually all of the airflow compressed by the forward fuselage during cruise conditions. At substantially slower speeds, the engines can still scoop up a significant fraction of the compressed air flow.

The airframe afterbody is also integrated with the engines, with the aft undersurface shaped to act as the upper side of a two-dimensional expansion nozzle. According to computer calculations performed at NASA's Langley Research Center, this surface would generate some 50% of the net thrust. The shock wave trailing back from the lower engine cowling would serve, in effect, as the lower side of the expansion nozzle.

Figure 7 shows some of the internal details of a scramjet engine module. The airflow, which has already been compressed in the vertical direction by the forward section of the fuselage, is now compressed laterally by wedge-shaped sidewalls and internal struts. Fuel is injected through multiple holes just behind the narrowest point of the passage through the engine, actively cooling the sidewalls and struts (particularly if liquid hydrogen is used as the principal fuel). Because the airflow has been compressed in only one direction at a time, it should be relatively easy to compute the airflow through alternative engine designs and to verify uniformity of the air/fuel mixture by suitable positioning of the fuel injection holes.

Since the airflow has been compressed 20- or 30-fold, the temperature of the air/fuel mixture exceeds the ignition temperature, even if the flow through the combustion section is only transonic.

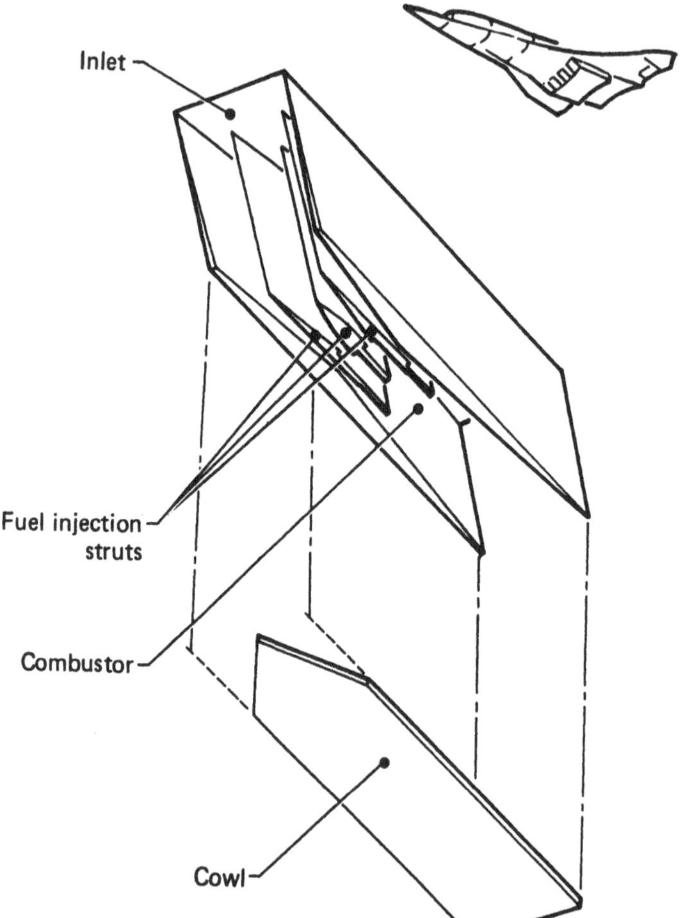

Figure 7. Internal details of an airframe-integrated scramjet engine module, with the lower engine cowl removed. Reprinted with permission from the American Institute of Aeronautics and Astronautics, Astronautics & Aeronautics, February 1978, pp. 38-48.

Thus, no glow plugs or other ignition devices should be necessary. Significant research work remains to be done, however, on the combustion processes in scramjet engines in order to assure stability of combustion and to understand how to optimize engine performance under the wide range of conditions a hypersonic transport will actually experience in routine flight operations.

Thus far, one of the best research tools available for combustion science is the shock tube. A long tube containing a uniform, prepared mixture of fuel, oxidizer, and inert gases (in whatever proportions are selected by the experimenter) is subjected to a shock wave produced by rupture of a membrane at one end of the tube. The shock wave heats the gas by rapid compression, and some time after the shock wave has passed a small element of the mixed gases, the mixture ignites. The delay between shock passage and actual ignition depends on the temperature of the mixture just behind the shock (which depends, in turn, on the strength of the shock) and on the composition of the mixture. Typically, for simple hydrocarbon fuels mixed with oxygen and argon, the ignition delay times can range from 0.1 to 100 micro-seconds as the initial post-shock temperature changes from 1000° C to 2200°C. If the airflow through the combustor in a scramjet has a speed of Mach 3, then one microsecond of travel time corresponds to one millimeter of displacement along the length of the engine. The flame front in a scramjet could thus vary in position by some 10 cm in a total engine length of 200-300 cm. Small fluctuations in inlet air densities and fuel injection rates can result in significant differences in combustion conditions and hence in combustion products and efficiency.

Small variations in the non-equilibrium conditions immediately behind the shock wave (during the first nanosecond, in some cases) may have major effects on the branches ratios for competing combustion reaction pathways for such simple fuels as methane, for which not even the equilibrium reaction rate with pure oxygen has ever been determined. During the ignition delay time, a host of chemical and pyrolytic reactions take place, producing a wide variety of intermediate species and radicals. This holds even for such a simple fuel as hydrogen in air: intermediate species include ozone, various oxides of nitrogen, hydroxyl radicals (OH), hydroperoxyl radicals (HO_2), and hydrogen peroxide. Clearly, the details of the combustion processes will have a major effect on the engine exhaust products. Far more research needs to be done on the details of combustion in any engine (automobile and aircraft piston engines, turbine engines, and ramjets included), but this is especially true for the scramjet with which we have as yet had so little pragmatic experience.

Clearly, development of the scramjet engine to a level comparable to the state-of-the-art for conventional turbine engines will be a lengthy and challenging effort. Including extensive wind tunnel testing and in-flight tests aboard high speed testbed vehicles, the development phase for the scramjet engine, if adequately funded, can be expected to last some ten to fifteen years. Major advances in numerical simulation methods for two- and three-dimensional hydrodynamics can be expected in the next decade (especially using finite element methods with moving node points), and these will aid enormously in such a program. Once the details of the combustion process in the engines are understood, a further difficult challenge will be to understand the effects of the exhaust products on the upper atmosphere, from the tropopause up to the cruising altitudes for a hypersonic transport.

Further development work will be necessary in materials science and in structural engineering to provide thermal insulation and active cooling for the aircraft interior against aerodynamic heating during cruise, since the wing area, dry weight, and fuel consumption can all be reduced by flying at the lowest practical altitudes which implies highest skin temperatures. Experience with the experimental rocket airplanes and with the Space Shuttle development program already provide a strong basis for flight controls at hypersonic speeds; some modest efforts to further improve our knowledge in this area would obviously be worthwhile in a hypersonic transport development program.

It is clear from the controversies over the SST, however, that a hypersonic transport would probably be prohibited from flight at supersonic and hypersonic speeds over land areas. Such restrictions on the HST would imply a significant decrease in capability from the travel times suggested in Figure 2. Some hope can be held out that such a ban might not be applied as widely against the HST because the sonic boom would be much more attenuated than that of an SST, due to the much higher cruising altitude of the HST. This question probably cannot be resolved satisfactorily before flight tests of an HST prototype. Such major uncertainties presently make the modest costs (perhaps a few billion dollars) of developing an HST highly unpalatable for the private sector. A variety of military missions for hypersonic airplanes (such as advanced reconnaissance aircraft, interceptor aircraft, strategic cruise aircraft and missiles, and highly maneuverable interceptor missiles) may well justify such a program in the public sector.

Boost-Glide Vehicles

As remarked earlier, a rocket engine is intrinsically far simpler than a scramjet or a conventional turbine engine, since the flow of the fuel/oxidizer mixture and of the combustion products can be sustained at a steady level by well-designed pumps and since the flow inside the combustion chamber is everywhere subsonic. Thus, it would appear plausible that rocket-powered flight in the atmosphere could be developed rapidly and economically. Rocket engines could be used to boost an aircraft to a very high altitude (perhaps 100 to 200 km), after which the aircraft would glide to its destination at high speeds, using conventional airbreathing engines for the last 10 to 30 minutes of flight for greater ease in integrating such a vehicle into other air traffic approaching the destination airport. The speed of the vehicle at its peak altitude could approach orbital speed, 25,000 miles per hour, permitting semi-global travel times as short as 90 to 120 minutes.[2]

Such vehicles, however, would be subjected to very high aerodynamic heating loads during glide for periods of an hour or more, with no fuel flow available to provide active cooling, since virtually all propellants would be consumed during boost. Westbound flights (launched against the direction of the earth's rotation) would be subject to significant reductions in payload for any given range of travel. Development of such vehicles could indeed be possible by the early 1990's for a cost of several billion dollars, perhaps as much as twice the cost of a hypersonic transport development program. As we shall see below, however, by spending slightly more money in the development phase, a much more versatile and useful vehicle could be developed in the same time-frame, namely an orbital transport vehicle. It thus appears that boost-glide transports are most unlikely to be developed in the near future.

Orbital Transports

Continuing advances in rocket engine performance, in advanced structural designs, and in higher strength materials promise major advances in launch vehicles capable of attaining low-earth-orbit (roughly 200 to 1000 km altitude). A number of conceptual design studies for advanced space shuttles show the feasibility of developing a single-stage-to-orbit (SSTO) vehicle by the early 1990's. In contrast to the Space Shuttle, which requires solid propellant rocket boosters and a throwaway external tank for its cryogenic liquid propellants, such an SSTO would be completely self-contained, permitting rapid turnaround between re-entry and landing and launch into orbit once again.

During the 1980's NASA plans a wide variety of experiments in orbit using the Space Shuttle to develop techniques for building large structures (up to several km across) in space; to develop new products which can be made in the weightless environment of orbit; and to explore the feasibility of collecting solar energy in space (where it is available 24 hours a day, unattenuated by the atmosphere) for transmission to the ground. Such projects and programs, included in the phrase "space industrialization," promise major commercial profits in the latter part of the 1980's, with rapid growth thereafter well into the next century. Since the useful lifetime of each Space Shuttle is expected to be about 100 flights, it is very likely that all four (or five) of the Space Shuttles presently planned will be worn out by 1990. Each Shuttle flight is expected to cost approximately $21 million to launch a payload of up to 32 tons, for a price per pound of payload around $325.

Present estimates for an SSTO capable of placing 250 tons into orbit suggest a launch price around $8 per pound of payload or about $4 million per launch (1978 dollars). Smaller SSTO's, more suitable for carrying large numbers of space workers and priority cargoes to and from orbit, could be expected to cost perhaps $20 per pound of payload. The R&D costs for a 10 to 12 year long program to develop such a family of vehicles would be $8 to $12 billion (1978 dollars), assuming the program starts in 1980 or 1981.[10] During the 1980's, NASA expects to launch as many as 900 Space Shuttle flights. The R&D costs would thus be amply justified by the difference in launch costs between the Shuttle ($325 per pound) and the SSTO ($20 per pound or less) during just 500 Shuttle launches.

One of the most venturesome programs of space industrialization presently under study by NASA, the Department of Energy, the aerospace companies, and numerous individuals in industry and academia is the concept of Solar Power Satellites (SPS), which were discussed in the 1978 Macro-Engineering Symposium.[11] Depending on how rapidly an SPS program were to be implemented, the total launch requirements during the mid- to late-1990's could reach several hundred thousand tons of payload per year, requiring up to 4 or 5 flights daily by a launch vehicle capable of delivering 250 tons into orbit. Each power satellite (with an electrical generating capacity of 10 million kw, assembled in space in 6 to 12 months, with a total mass of about 100,000 tons) would provide some $3.2 billion a year of revenues from sale of electrical power, so the economic motivations for development of the SSTO in connection with an SPS program are very strong.

Less ambitious space industrialization programs involving advanced communications, navigation, datalink services, and earth resource prospecting satellites would also justify the development of SSTO vehicles. Construction and maintenance of very large antenna arrays in geosynchronous orbit would facilitate worldwide portable personal telephone service (a Dick Tracy-style wrist telephone

	747 Airliner (HTOHL)	NASA baseline shuttle (VTOHL)	Boost-glide rocket transport[a] (VTOHL) Ref 14	Orbital rocket transports[a] (VTOVL) Ref 15	(HTOHL)[b], Ref 16	(VTOHL)[c], Ref 17	
	JP-4	Solid + O_2/H_2 2-1/2-stage	O_2/H_2	O_2/H_2 H_2 tanks retained	O_2/H_2 + ground accelerator	$O_2/HC/H_2$ 2 dual-fuel engines	
						$O_2/RP\text{-}1/H_2$	$O_2/RJ\text{-}5/H_2$
GLOW, lb	713,000	4,101,000	3,490,000	3,336,000	3,260,000	4,450,000	5,000,000
Thrust, lb	188,000	7,000,000	4,360,000	4,150,000	6,000,000 (incl. accel.)	6,050,000	6,800,000
Pr. vol., ft^3	5,200	90,000	132,000	128,000	127,000	83,000	83,000
Dry wt., lb	353,000	540,000	470,000	315,000	405,000	372,000	380,000
Payload, lb[d]	103,170	52,000[e]	60,000	50,000[e]	60,000[e]	65,000[e]	85,000[e]
(or No. pass.)	374	—	200	170	200	200	250

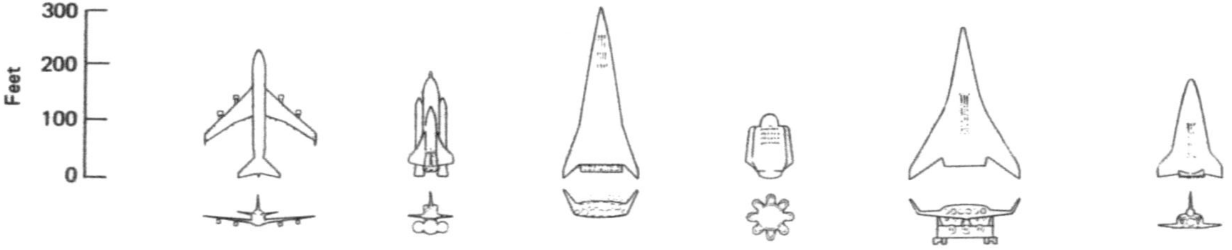

a. Designs & performance not optimized
b. Adapted & scaled from Ref 16
c. Adapted from Ref 17
d. 2-G acceleration limit
e. 100 n. mi./60° orbit

Figure 8. Comparison of boost-glide and orbital rocket transports to B-747 airliner and NASA Space Shuttle. Adapted from Robert Salkeld, "Global Rocket Transports: Changing Perspectives," in The Second Fifteen Years in Space, Vol. 31, Science and Technology, American Astronautical Society, San Diego, 1973.

capable of global access) and other advanced information handling capabilities.[12] The total launch requirements for such systems may total only a few thousand tons each year, as well as a few dozen to a few hundred workers transported to and from orbit each year (rather than several thousand workers as in the case of an SPS program). Yet revenues from such systems may easily total $20 to $50 billion per year by the year 2000, growing to several hundred billion dollars annually by 2010. Given such lucrative industries in space, the development of an inexpensive SSTO by the 1990's is virtually inevitable.[13]

If the price per pound for launch into orbit is $10 to $20, then the cost for a passenger ticket into space would be between $3,000 and $10,000, depending on crew size, load factors, utilization rates, and degree of personal amenities provided for each seat. But the capability of carrying a passenger into orbit in a shuttle which can land anywhere on earth implies a global transportation capability. The pricetag for a trip half-way around the world would be about the same as the price to go into orbit, and that price is reasonable enough for some of the high-priority travel needs I mentioned at the beginning, especially when combined with a travel time of two hours or less between any two points on earth.

Figure 8 compares several SSTO concepts to the Boeing 747 airliner, to the Space Shuttle of the 1980's and to a boost-glide rocket transport. Each vehicle shown is characterized by its takeoff/landing operating mode, propellants used, gross liftoff weight (GLOW), total thrust of all engines, propellant volume, dry weight, and payload (or passenger complement). Payloads are based on performance capabilities subject to a peak acceleration of 2 g's; for orbital vehicles, a nominal 100 nautical mile altitude (184km) orbit inclined 60° to the equator has been assumed. (Because of the earth's eastward rotation, launching due east from the equator gives the best performance possible for a given vehicle. Launching into orbits with inclinations of less than 60°, even from moderate latitudes, will allow somewhat higher payloads than those shown in Figure 8.)

While rocket engines are far simpler than turbine engines or scramjets, they are also far noisier for comparable vehicles since their thrust must be at least an order of magnitude greater. Operation of rocket engines is thus not likely to be permitted close to densely populated areas. But if the advantages of short travel times are to be realized, an SSTO should be able to operate close to urban centers, at least on one end of the trip. Horizontal landing (HL) configurations (such as the Space Shuttle) can land at conventional airports with no more noise than conventional jet transports. Such configurations are ideally suited to aerodynamic reentry and glide, with conventional subsonic turbine engines used during the last ten to thirty minutes of flight to facilitate integration with other air traffic approaching the destination airport.

Horizontal takeoff (HTO), on the other hand, poses major engineering problems for SSTO's because of the enormous mass of propellants onboard before launch. This would require massive, sturdy landing gear which would be useless in orbit and far heavier than necessary for landing, when the vehicle's tanks are empty. The high weight at liftoff would also require takeoff speeds much higher than touchdown speeds, further aggravating the landing gear problems. (Takeoff speeds of about 400 mph would be required for wings designed for a landing speed of about 150 mph.) One solution to this dilemma is shown in Figure 8. The SSTO itself is equipped with lightweight landing gear adequate for touchdown with tanks empty. For takeoff, the SSTO is placed on top of a ground accelerator, a rocket-powered dolly with heavy rolling gear which stays on the runway after bringing the SSTO to liftoff speed. Such operations, however, seem difficult to reconcile with conventional airport operations, and would require far longer than normal runways.

Another approach (not shown in Figure 8) is to take off with only partially filled tanks on lightweight landing gear, rendezvous with a very large tanker, and fill the tanks in flight. Development of a massive tanker capable of transferring cryogenic propellants at ten times the rate presently used by military aircraft is not a simple matter, and would escalate development costs for the SSTO system significantly. As yet, however, we cannot rule out this concept. A third approach (not shown in Figure 8) is to carry the fully-fueled SSTO to high altitude with a large conventional jet aircraft and to launch the SSTO in flight.[12] Again this approach would complicate operations and add to the development costs of the system. Either of these approaches, however, would permit departure from close-in airports, with rocket ignition delayed until the SSTO is in a more remote area at stratospheric altitudes.

Vertical takeoff/vertical landing (VTOVL) is the operating mode familiar to fans of grade-B science fiction movies. Rocket ships re-entering from space come in tail first, igniting engines a few minutes before landing, to touch down on a large concrete blast pad amid flames and smoke. Such vehicles, having neither landing gear nor wings, would have somewhat better performance than Shuttle-type vehicles. On the other hand, if reentry is performed tail first, a more sophisticated type of rocket engine, the plug-nozzle rocket, must be developed, to permit closing the engine openings and to withstand the heat during reentry.

Further developments may make some of the above conclusions regarding HTOHL and VTOVL vehicles obsolete, since the field of SSTO concepts is evolving rapidly and vigorously. On the basis of published literature available at the present time, however, it appears that a vertical takeoff/horizontal landing (VTOHL) configuration is most likely for the next generation of earth-to-orbit vehicle, as shown in two alternate configurations. in Figure 8. Instead of a pure liquid oxygen/liquid hydrogen system, these configurations use dual fuel engines, burning hydrocarbon fuel first, then switching to liquid hydrogen. Liquid hydrogen is lighter than hydrocarbon fuels for the same energy content, but requires much more volume than hydrocarbon fuels of the same energy content. Thus, a rocket using only hydrogen would require much larger tanks and thus have a higher dry weight, as well as incurring greater air resistance during ascent. A rocket using only hydrocarbon fuel, on the other hand, would have a greater GLOW, reducing performance capabilities of the vehicle due to the additional propellants needed to lift the greater

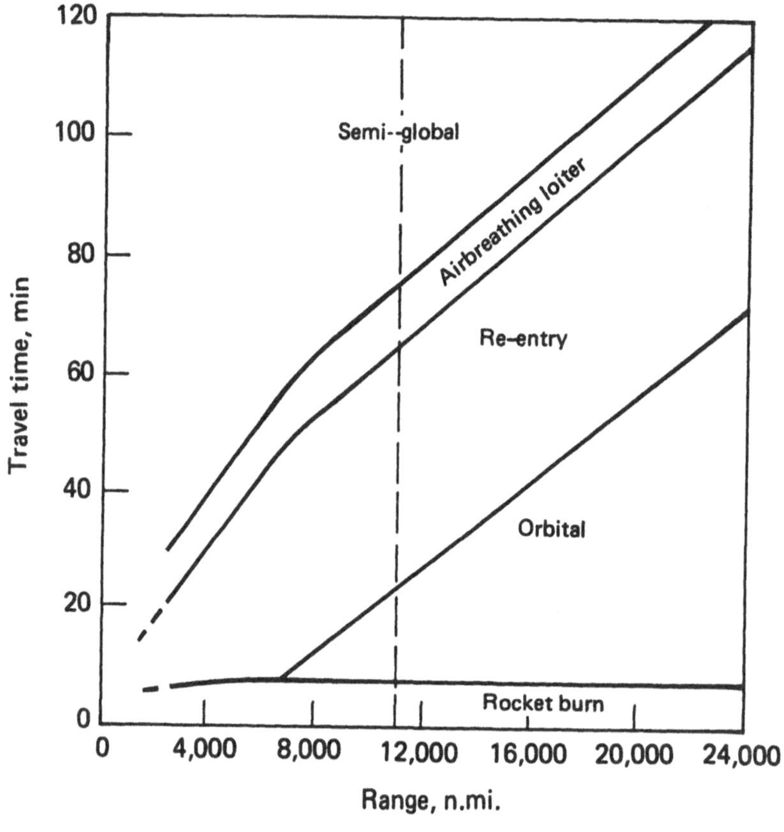

Figure 9. Flight profile for a global rocket transport. Adapted from Robert Salkeld, "Global Rocket Transports: Changing Perspectives," in The Second Fifteen Years in Space, Vol. 31, Science and Technology, American Astronautical Society, San Diego, 1973.

total weight initially. The optimum vehicle turns out to be a compromise between the two fuels, with the denser fuel burned off first.

While a VTOHL vehicle could not depart from a close-in airport, it can land close in and then ferry itself (using conventional turbine engines) to a more remote site for launch into space again. Figure 9 shows the flight profile for trips of various ranges. For destinations to the west of a given departure point, the SSTO would still be launched in a generally easterly direction and go around the long way. Trip ranges of up to 24,000 nautical miles are thus shown in the figure. For some pairs of cities, a second orbit may be necessary if launch performance (and payload) is not to be significantly degraded by using orbits with inclinations of more than 60°. This is most likely to happen for city pairs half-way around the globe in longitude, and on the same side of the equator. In such cases, one and a half orbits can do the trick, for a total trip time in these cases of about three hours from launch to landing.

The global transport rocket as described here would differ from a vehicle used strictly for earth-to-orbit transport primarily in the addition of conventional jet turbines and fuel tanks and the removal of extended orbital life-support systems and of orbital maneuvering rockets and propellant tanks. Little additional cost would be incurred in adapting an SSTO developed for space industrialization purposes into a global transport rocket. Depending on costs of hydrocarbon and hydrogen fuels in the 1990's (which no one can predict with high certainty), it seems that the ticket price for semi-global ranges by orbital flight would be about two to three times the cost of subsonic flights over the same range, entirely reasonable for a number of the high-priority travel needs discussed earlier. A fleet of perhaps 50 to 100 such vehicles should suffice for the anticipated priority travel needs around the turn of the century. The construction of launch facilities for a VTOHL vehicle at sites somewhat remote from urban areas would be comparable in cost to construction of new airports. Sonic booms during launch would be a minimal problem due to the steep ascent of a VTOHL, while deceleration during reentry and glide would minimize sonic booms to those generated at very high altitudes. If necessary, approach paths can be designed to minimize the impact of even these sonic booms, since the wings of such a vehicle combined with its conventional turbine engines would permit the vehicle to veer off as much as 1100 nautical miles from the extrapolated ground track of its orbit. (This lateral maneuvering distance is called "cross range.")

Comparative Assessment

While the global subway system and the laser-powered airplane system both appear to be economically viable in the sense that anticipated revenues could amortize the initial investments, plus interest charges, the initial investments involved are very large and thus imply a great deal of risk. Such projects can be financed in a variety of ways[19], but dedicated and ardent enthusiasts would be **necessary** prerequisites for any of these methods of financing very large projects of such scale, making these candidates for global transportation most unlikely during the next few decades.

The boost-glide rocket transport seems to be too limited in its capacity to justify its development costs, when a small increment in the R&D costs would achieve fully orbital capability. The sonic boom throughout its flight path would pose significant restrictions on routes and thus on its economic viability. All in all, it seems that the boost-glide concept will go down in history as an amusing evolutionary dead-end.

The hypersonic transport would require a substantial R&D investment (much less than the costs of global subway or laser-powered flight, but still large compared to aircraft developed thus far) and may have difficulty proving its economic viability. Should the R&D costs be largely assumed by the military for the purpose of developing hypersonic bombers, reconnaisance planes, or interceptors, however, an HST may emerge as a spin-off.

The single-state-to-orbit vehicle, it would seem, is the leading contender for rapid global transportation around the turn of the century, since its development costs will be readily supported by the requirements of highly profitable ventures in industrial operations in orbit beginning in the late 1980's. Use of a slightly modified SSTO for civilian global transport will be a low-cost spin-off of this development program, which is technologically feasible in 5 or 10 years less time than is development of a viable scramjet.

Nonetheless, a modest research and development program in hypersonic flight would be well worth continuing, since it would provide a concrete focus for research in a number of vitally important and valuable fields. Some of these include a basic scientific understanding of the microscopic details of combustion; advances in numerical modeling of coupled chemical kinetics/hydrodynamics/radiation transport; advanced aerodynamic shapes and controls; advanced high-strength and high-temperature materials; safe and practical use of cryogenic fuels in aviation; interactions of exhaust products (including water vapor) with the stratosphere and ionosphere; propagation of sonic booms generated at high altitudes; and advanced structural concepts for thermal barriers.

How big can an advanced technology be and still be beautiful? Global subways and laser-powered flight seem too big to be viable, unless motivated and driven by a still larger program for other purposes. Hypersonic flight, on the other hand, seems too small to fly--its prospective markets seem too marginal to safely justify the R&D costs, and its development timetable is too long compared to that for an SSTO. The single-stage-to-orbit, adapted to global transportation needs, seems just right to work.

References

1. Kenneth Clark, Civilization: A Personal View, Chap. 11, "The Worship of Nature," Harper and Row, New York, 1969.

2. Robert Salkeld, "Global Rocket Transports: Changing Perspectives," in The Second Fifteen Years in Space, Vol. 31, Science and Technology 1973, American Astronautical Society, Tarzana, Calif. 91356.

3. Robert M. Salter, "Transplanetary Subway Systems: A Burgeoning Capability," pp.98-129, in: Frank P. Davidson, L.J. Giacoletto, and Robert Salkeld, eds., Macro-Engineering and the Infrastructure of Tomorrow, AAAS Selected Symposium 23, Westview Press, Boulder, CO,1978.

4. F. Chilton, B. Hibbs, H. Kolm, G.K. O'Neill, and J. Phillips, "Electromagnetic Mass Drivers," in G.K. O'Neill, ed., Space Manufacturing from Non-Terrestrial Materials, Vol. 37, Progress in Aeronautics and Astronautics, American Institute of Aeronautics and Astronautics, New York, 1977.

5. William T. Hamilton, "Projected Aircraft Systems Development," paper AAS 78-194, American Astronautical Society, San Diego, Calif. 1978.

6. I. Bekey, H. Mayer, and M. Wolfe, Advanced Space System Concepts and Their Orbital Support Needs, final report for NASA Contract NAS-W-2727, Aerospace Corp. Dec., 1976.

L.N. Myrabo, "Solar Powered Global Aerospace Transportation," AIAA Paper 78-689, IAAA/DGLR 13th International Electric Propulsion Conference, American Institute of Aeronautics and Astronautics, New York, 1978.

A. Hertzberg, and K. Sun, "Laser Aircraft Propulsion," in K. Billman, ed., Third NASA Conference on Radiation Energy Conversion, Vol.61, Progress in Aeronautics and Astronautics, American Astronautical Society, San Diego, CA, 1978

W.S. Jones, "Laser Powered Aircraft and Rocket Systems with Laser Energy Relay Units," in K. Billman, ed., Third NASA Conference on Radiation Energy Conversion, Vol. 61, Progress in Aeronautics and Astronautics, American Astronautical Society, San Diego, CA., 1978.

7. A.R. Kantrowitz and R.J. Rosa, "Ram Jet Powered by Laser Beam," U.S. Patent No. 3,818, 700.

8. K. Sun, A. Hertzberg, and W.S. Jones, "Laser Aircraft Propulsion," Paper AAS 78-157, American Astronautical Society, San Diego, Calif., 1978.

9. Robert A. Jones and Paul W. Huber, "Toward Scramjet Aircraft," Astronautics and Aeronautics, pp. 38-48, February, 1978.

10. B.Z. Henry and C.H. Eldred, "Advanced Technology and Future Earth-to-Orbit Transportation Systems," in: Jerry Grey, ed., Space Manufacturing Facilities-II, American Institute of Aeronautics and Astronautics, New York, 1978.

11. Peter E. Glaser, "Solar Power Satellites: The Promise and Challenges," in: Frank P. Davidson, L.J. Giacoletto, and Robert Salkeld, eds.,Macro-Engineering and the Infrastructure of Tomorrow, AAAS Selected Symposium 23, Westview Press, Boulder, CO., 1978.

12. I. Bekey, H. Mayer, and M. Wolfe, "Advanced Space Systems Concepts and Their Orbital Support Needs," The Aerospace Corp. Report No. ATR-76(7635)-1, NASA Contract NAS-W-2727, El Segundo, Calif., 1976.

B.I. Edelson and W.L. Morgan, "Orbital Antenna Farms," Astronautics and Aeronautics, p. 20, September, 1977.

13. "Space Industrialization 1980-2010," Final Report, NASA Contract NAS 8-32197, Science Applications, Inc., Huntsville, AL, April 1978.

14. M.W. Hunter, II, and D.W. Fellenz, "The Hypersonic Transport: The Technology and the Potential," AIAA Paper 70-1218, American

Institute of Aeronautics and Astronautics, New York, 1970.

15. P. Bono, "Pegasus: A Design Concept for a V.I.P. Orbital/Global Rocket Transport," SAE Paper No. 64087, SAE National Aeronautics and Space Engineering Meeting, Los Angeles, CA., Oct. 7, 1964.

16. "Feasibility Study of Single-Stage-to-Orbit Vehicle," Final Report, The Boeing Company, Air Force Contract F04701-72-C-0418, Seattle, WA, January, 1973.

17. Robert Salkeld, "Mixed-Mode Propulsion for the Space Shuttle," Astronautics and Aeronautics, pp. 52-58, August 1971.

    Robert Salkeld and R. Beichel, "Reusable One-Stage-to-Orbit Shuttles: Brightening Prospects" Astronautics and Aeronautics, June, 1973.

18. Robert Salkeld and Robert S. Skulsky, "Air Launch for Space Shuttles," Acta Astronautica 2 pp. 703-713 (1975).

Section C

Solar Power Satellites and Automation in Space

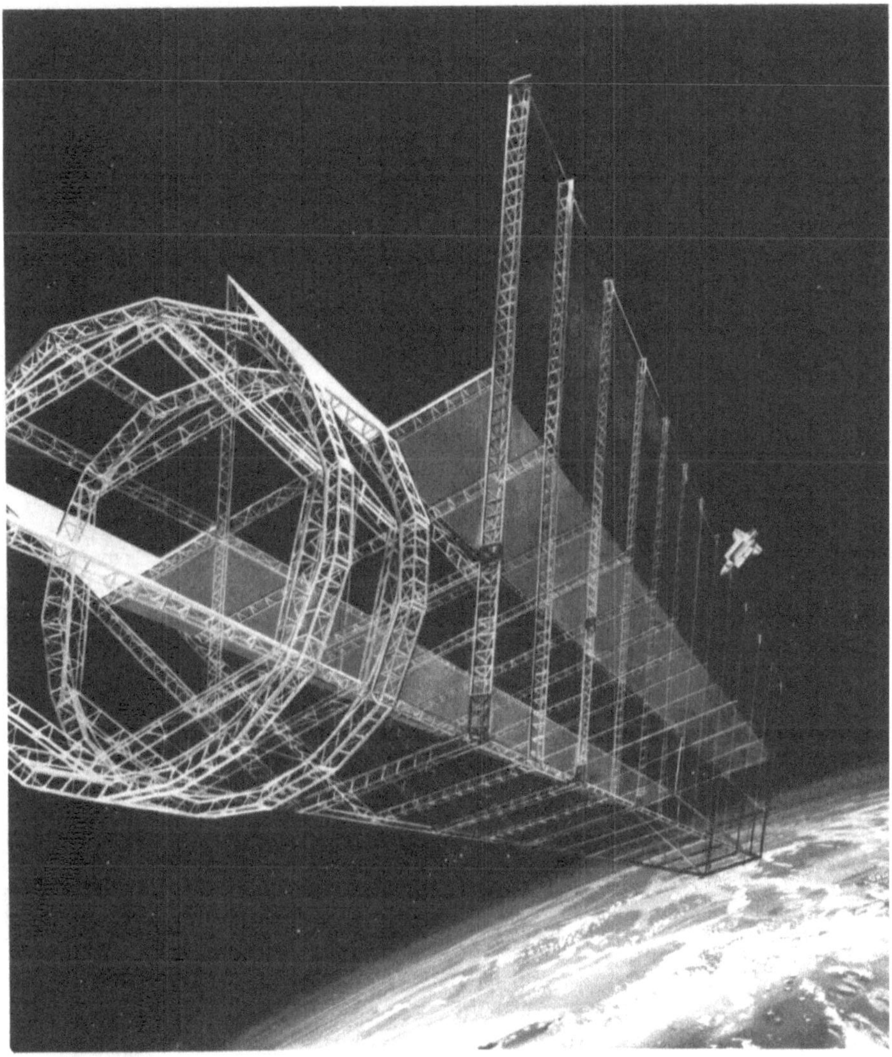

Space Shuttle can deliver both the materials and the machinery required to build large space structures, such as this demonstration satellite solar power station. After being fabricated and assembled in low earth orbit, a newer station would be transferred to its permanent place in geosynchronous orbit (about 22,000 miles out in space) where it would beam a continuous stream of microwave energy to earth receivers, which would convert the energy to electricity. Courtesy of Rockwell International, Space Division.

Overview

SOLARES is a space-terrestrial solar energy system consisting of a set of orbiting mirrors that would provide nearly continuous reflected sunlight to a world-distributed set of solar conversion sites. Kenneth W. Billman, William P. Gilbreath, and Stuart W. Bowen argue that this system promises renewable solar energy which will be economically competitive with alternative fossil and nuclear sources. Related studies are cited to support the technical and economic feasibility of the SOLARES system. Questions of environmental impact and social-political acceptance remain to be answered.

Kenneth W. Billman,
William P. Gilbreath and Stuart W. Bowen

15. Solar Energy Economics: Orbiting Reflectors for World Energy

Introduction

The natural storage processes for solar energy have, over the eons, produced an inexpensive supply of fossil fuels which has prevented the economic acceptability of direct solar energy use. Today, when these reserves of stored solar energy are perceived to be inadequate for future generations, when their increased cost ferments political and economic instability, and finally, when the quality of life can be degraded by their mining, transport and burning, the search for economic methods of using the "free" and perpetually available solar energy source has intensified.

This paper outlines the recent study we have made on a space-terrestrial solar energy system, which we call SOLARES, consisting of a set of orbiting mirrors that provide nearly continuous reflected sunlight to a world-distributed set of solar conversion sites. As will be shown, this added insolation dramatically reduces the cost of direct solar-electric production. Additionally, the magnitude of energy supply on a global scale can be substantial. This combination of significant and renewable energy supply to the world community, economically competitive to alternative fossil and nuclear sources, is the promise of the SOLARES concept.

Accompanying this promise, there are, of course, many challenges which must be met. Some have been identified in our technical assessment. Others will certainly emerge if SOLARES becomes an accepted NASA program and in-depth engineering,

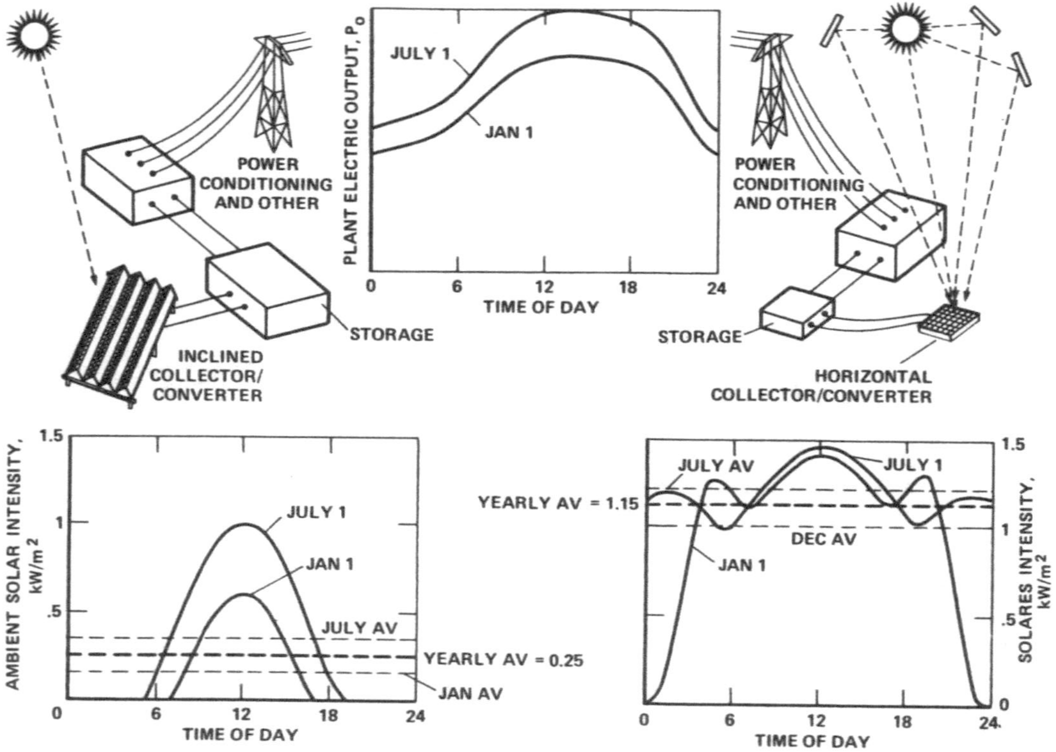

Figure 1. An illustration showing the decreased collector/converter, storage and other costs of a solar farm resulting from orbiting mirror insolation.

systems, and economic analyses are performed.

Without these in hand, it may appear presumptuous to accord SOLARES the promise of significant economic improvement to solar-electric production. However, as will be seen, the predominant cost element of the system is the solar terrestrial converters. Fortunately, this element has been intensively studied by others, for ambient sunlight use, and firm engineering and circa 1990 costs have been established. The secondary SOLARES cost element is presently believed to be the space system--resulting from the simplicity, efficiency, and especially the low mass density of the passive solar reflectors which we, and others, envision as suitable. It is in this space element where one could anticipate sizable change with deeper analysis. But, as will be shown, even five-to-ten-fold increases in its cost will still allow SOLARES to be competitive to alternative coal, oil, and nuclear electric power sources! It is this insensitivity to space costs, coupled with the firmer terrestrial costs -- which, by scientific analysis, can be demonstrated to be significantly reduced with the SOLARES approach -- that lends credence to our assessment of the importance of this new energy alternative.

The impact of SOLARES on the economics of solar-electric generation can be seen in Figure 1. We consider here two solar farms: one operating with normal sunlight, the other with normal sunlight augmented with space mirror-reflected insolation. If these systems are to compete with alternatives, they must produce baseload (continuous) power. Typical power output or "demand" curves are shown for 24-hour periods for summer and winter. To meet these, each system must provide expensive energy storage. From the normal solar insolation curve (lower left) it can be seen that this is substantial: demanding about 14 hours of capacity. It is for this reason that economics normally preclude a baseload solar energy system. Careful analysis of the SOLARES insolation curves show that although some eclipsing of the mirrors by the earth occurs around local midnight at the site, the required storage is reduced typically by a factor of five. Secondly, the "diluteness" or low average energy density of ambient sunlight is noted. The yearly average for a good solar farm site is only about 0.25 kW/m^2, with a two-fold seasonal variation.

This low density demands an extremely large area of expensive solar collectors and converters for significant energy output. The use of concentrators to reduce the area of expensive converters required is unfortunately accompanied by the need for tracking the sun which results in an even more costly and complicated ground site. In contrast, the SOLARES mirror system is sized to provide a yearly average 1.14 kW/m^2 (typical high-noon desert intensity), with a significantly lower seasonal variation of about $\pm$ 10%. Hence, to produce the same electric output, the SOLARES solar conversion array is reduced by a factor of $1.14/0.25 \simeq 4.6$.

This potential reduction of nearly 5 in each of the most costly items of solar farming--storage and solar converters--constitutes the promise of SOLARES. The challenge, in the main, is to accomplish this by means of a space system whose cost does not erode this economic advantage. It is our belief that our design of the SOLARES mirror system which makes extensive use of the thin film, large area structures which appear possible in the weightlessness of space, the use of solar sailing to reduce space transport costs and other innovations should meet this challenge.

This report will address the SOLARES concept under those four qualities which must be possessed by any new energy alternative in order to achieve acceptance: TECHNICAL FEASIBILITY, SIGNIFICANT AND RENEWABLE ENERGY IMPACT, ECONOMIC FEASIBILITY, AND SOCIAL/POLITICAL ACCEPTABILITY.

Technical Feasibility - A SOLARES Challenge

The initial technical assessment of the SOLARES concept[1], and subsequent detailed study[2-3] have evolved a space-terrestrial system which, in principle, will work. As will be seen, however, its requisite magnitude -- to make a significant impact on world energy needs -- is large and has few parallels in history. Fortunately, the system consists of highly redundant space and terrestrial elements, amenable to the economics of mass production and most importantly, capable of early and decisive proof-of-principal testing with the Space Shuttle and in normal solar conversion simulations. We review here the technical aspects of these space and terrestrial components.

The Space System

An artist's sketch of the orbiting mirror system as seen from space, Figure 2, illustrates the method of providing nearly continuous, enhanced solar energy to a typical ground site. Each mirror orbiting the earth with a particular period, is continually oriented when over a site to reflect its solar image onto a common ground spot. The image size is prescribed by geometric optics, having a minimal circular diameter of $D_M + h\alpha$, where D_M is the mirror diameter, h its altitude above the earth's surface, and α is the subtense angle of the sun (0.0093 rad $\simeq$ 1/100 rad). For a non-zenith mirror, this image is increased in size and is elliptical in shape. The superposition of such intensity profiles from all the mirrors over the site constitutes the SOLARES mirror contribution to the conversion farm; additionally, during the daylight hours, the ambient or direct sunlight is used. Detailed numerical modeling has shown that the terrestrial intensity profile resulting from all mirrors above 30° from the horizon, the "useful viewing circle" or UVC, is relatively uniform over a circle of diameter $h\alpha \simeq h/100$ and diminishes rapidly beyond this. This "$h\alpha$-circle" thus sets the scale of a SOLARES conversion farm and, interestingly, also sets the scale for the total mirror area needed in the UVC to provide a given value of average radiant energy density delivered to the site. If the latter is chosen as 1 kW/m^2 the oversite mirror area is found roughly equal to 3 times the $h\alpha$-circle area. This leads to certain important conclusions. First, the use of a single, geostationary mirror which would remain fixed in position above a single site appears excessive in land and mirror areas required since the orbit altitude of 35,800 km would give a spot size of about 333 km or the size of the state of Maine! Lower orbits ranging in value from the lower limit of about 1,000 km (period $\simeq$ 2 hr), dictated by atmospheric drag on the low ballistic coefficient mirrors, to an upper limit of about 6,500 km (period $\simeq$ 4 hr), dictated by the predicted world energy needs circa 1990, are of more significance. The second resulting conclusion is illustrated in Figure 3, viz. a world-wide orbital mirror set is necessary so that as some mirrors leave the UVC of a given site, others enter it and hence an uninterrupted energy supply is maintained. The added

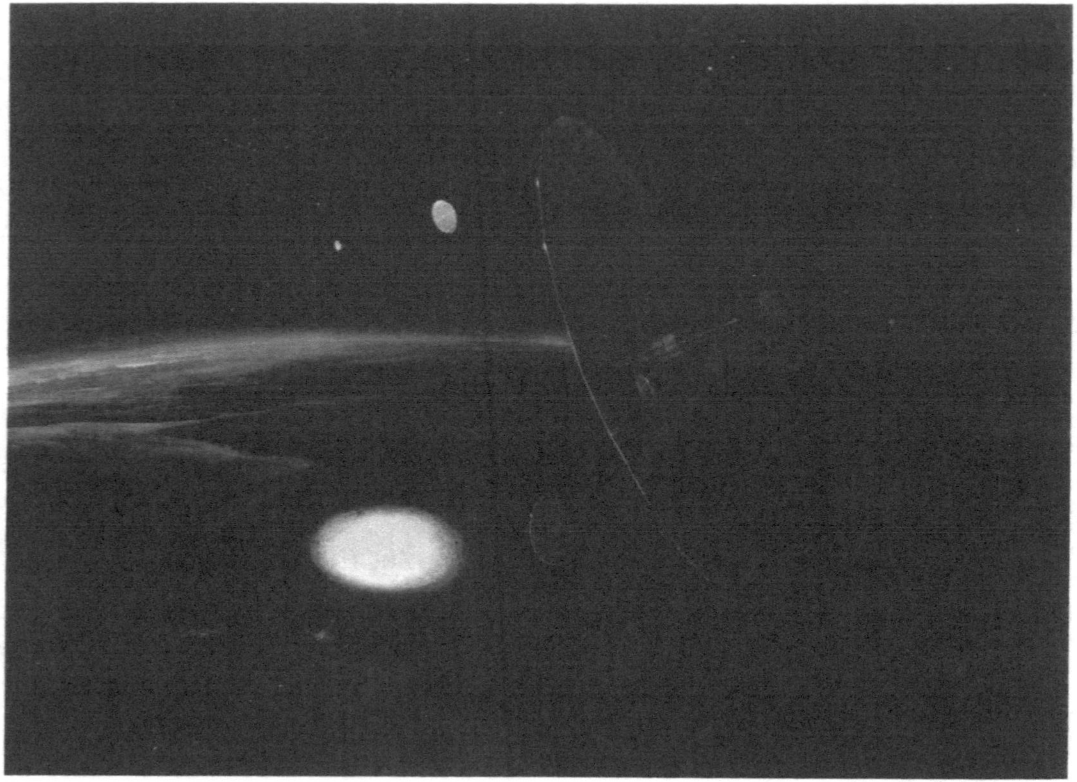

Figure 2. Artist's concept of a space view of the orbiting mirror system.

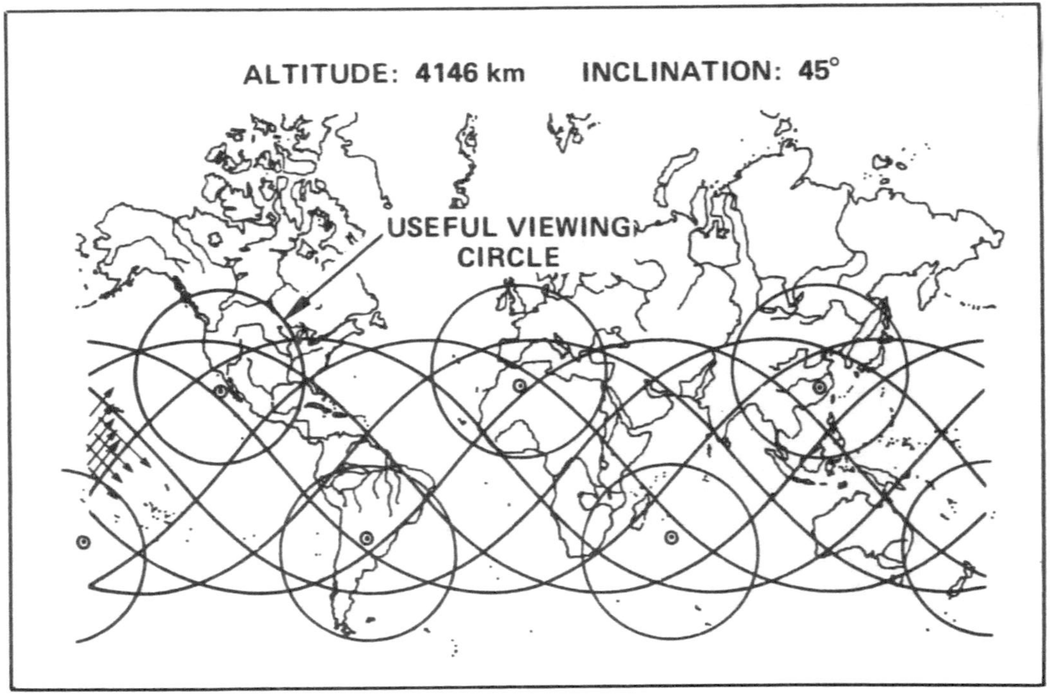

Figure 3. Ground tracks of a typical mirror showing its daily coverage of the Earth. Other mirrors follow similar, interleaved orbits.

complexity of such a multiple mirror set over that of a single mirror is believed to be minimal since all such mirrors are "free-flyers", that is, independent and individually controlled in either case. Additionally, many sites may now be serviced simultaneously. The maximum number is fixed by choice of UVC size, depending on altitude and the chosen acceptance angle, and the allowable reorientation torques which may be applied to the low-mass mirror structure. These are necessary, of course, to prepare a mirror for service to a site during the time it spends orbiting between it and the prior site. This allowable site number ranges between 16, for low altitude mirrors (1196 km) serving equatorial sites, and 4, for high altitudes (6384 km); the former would produce electric power at a level of about 13 GWe/site while the latter, larger sites would deliver about 375 GWe each. For reference, current usage in the United States is about 250 GWe. It is interesting to note that the SOLARES system can be expanded, to meet increasing world energy demand, by raising the altitude of the mirrors and simultaneously increasing the conversion site radius to accommodate the resulting larger hα-circle. Such a growth scenario will be explored later in this report.

Orbits

Many special orbit types, such as geostationary, polar and sun-synchronous, have been considered; each possesses certain advantages and disadvantages. Of most utility are circular orbits of inclination somewhat larger than that of the latitude of the northernmost (or southernmost) sites to be serviced. We exclusively consider here two particular cases: those at about 40° inclination, servicing sites at 30° latitude and those at 0° servicing equatorial sites. The latter, equatorial orbits, are of particular interest since they make most effective use of the mirrors -- each mirror passes over each equatorial site on each successive orbit. Hence, these require a smaller mirror set (or space expense) to deliver the prescribed insolation.

Insolation

To quantitatively determine the mirror system reflective area required, extensive numerical computation has been made. The physical model used[3] places mirrors of unit area into prescribed orbital positions over the site and sums the transmitted intensities of each to the prescribed (latitude and longitude) site for each incremental time step during the 24 hours of the particular date being examined. All geometric factors, the equation of time, seasonal variation of the effective solar constant, eclipsing (earth shadowing) of certain mirrors, and the absorption and scattering by the atmosphere are rigorously included. Typical outputs from these computations are illustrated in Figure 4 for the case of a 30° latitude site and mirrors at 6384 km altitude with 40° inclination. On the left is seen the spatial radiant intensity profile, showing the uniformly illuminated hα-circle with rather sharp decrease in intensity outside it. On the right, two temporal plots illustrate the variation of the value of this central hα-circle intensity during typical summer and winter days. Ambient sunlight is also shown. Interestingly, it is seen that although the insolation provided by the mirrors decreases around noon, since the effective mirror area is geometrically lessened at this time, the ambient solar insolation is large and hence the total insolation available to a horizontal ("flat-plate") conversion system is reasonably constant. The effect of eclipsing during the winter months is seen occurring around local midnight in the lower curve. As discussed earlier, minimal storage (cross-hatched area above the dashed demand curve) suffices to provide continuous baseload output during these periods (cross-hatched area under the demand curve).

From these insolation curves, together with the determination of the fraction of the space system mirror area which is above the site, an accurate assessment of the total space system mirror areas required to give a yearly averaged site radiant intensity of 1.14 kW/m^2 is determined. As seen in Table 1 for various inclined and equatorial orbits, the total areas are large, ranging from about 6,000 km^2 to 70,000 km^2. If 1 km^2 mirrors were used, this would dictate, of course, a similar number of mirror satellites. Indeed, if, as will be shown, this system can nearly duplicate in output that of the existing world system of hydro-, fossil- and nuclear-electric power plants, it is not surprising that a large system is

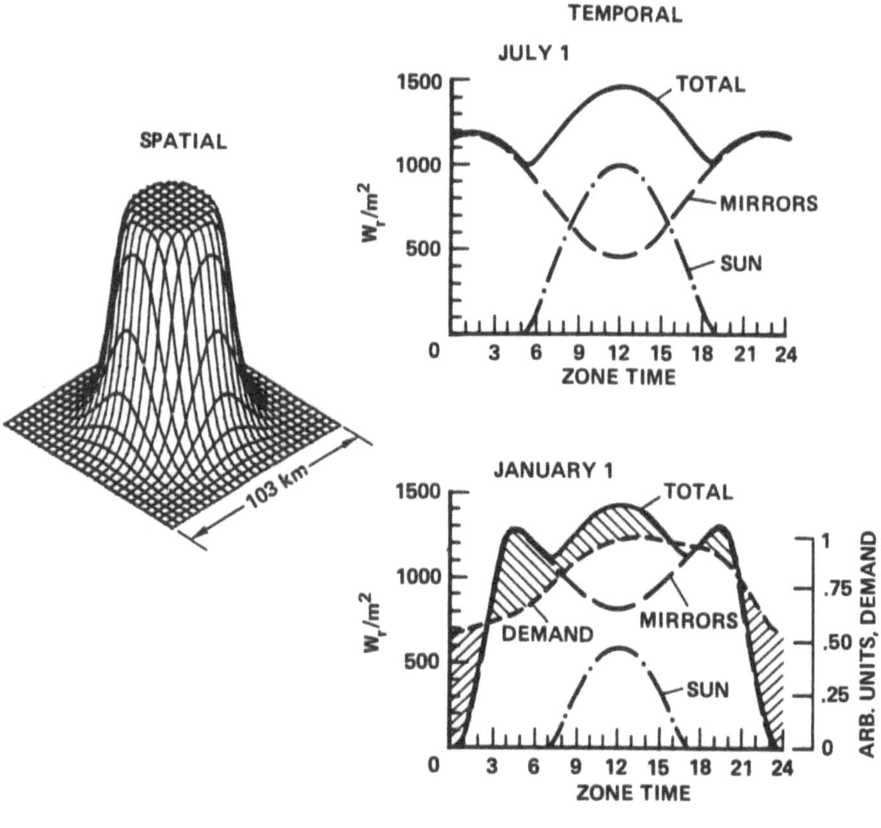

Figure 4. Spatial insolation profile (left) and intensities at site center for typical summer and winter days.

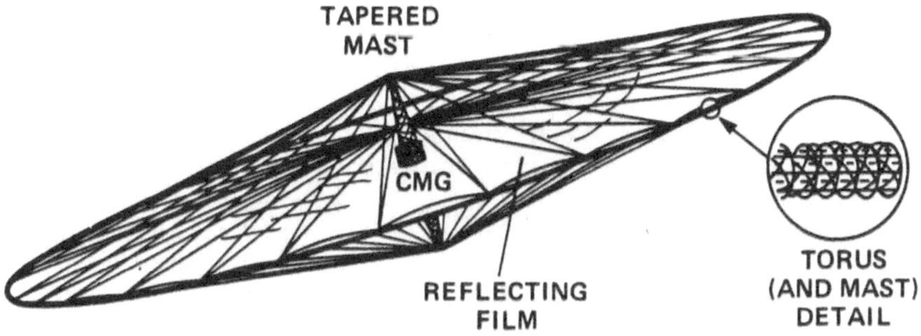

Figure 5. Illustration of the Ames SOLARES mirror configuration.

Table 1. Total space system mirror areas required to provide a nominal 1.14kW/m^2 on sites[a].

Mirror altitude, 10^3 km	Orbit inclination, deg	Orbit period, hr	Total mirror area, 10^3 km^2	$h\alpha$-circle diameter, km
1.18	0 (Equatorial)	1.82	6.4	10.95
2.10	0 (Equatorial)	2.16	10.4	19.52
4.14	0 (Equatorial)	2.98	21.6	38.5
6.38	0 (Equatorial)	3.98	36.4	59.34
1.20	45	1.82	29.1[b]	11.12
2.11	40	2.16	25.8	19.61
4.14	40	2.98	45.8	38.56
6.38	40	3.98	69.0	59.38

[a]Includes ambient insolation.

[b]Eclipsing raises this area above that for the 2110 km orbit.

necessary. The important question is not scale, but rather those of technical feasibility and economic desirability of achieving it. Both, we believe, can be answered affirmatively.

Mirror Design

Of critical importance to minimization of the space cost component of SOLARES is to reduce the mass which must be placed into orbit. Since the mirror areas are specified, the areal density of these structures must therefore be minimized. Fortunately, the space environment is favorable to this since an orbiting object is weightless and, at sufficient altitude, there is no structure loading by atmospheric viscous drag. These factors point to the possibility of very large area (1-50 km^2) low average mass density (10-200 gm/m^2) reflectors which can maintain their required optical surface flatness under the residual forces: radiation pressure, gravity gradients, and the inertial loading due to intra-and inter-site orientational torques.

Our "Mark I" design of such a structure is schematically shown in Figure 5. It consists of a thin (ca. 2 μm) polymeric film, overcoated (ca. 0.1 μm) on each side with highly reflective aluminum, tautly supported by a bicycle-like support structure consisting of an outer toroidal rim, a tapered mast (or hub) and guy wires (spokes) between these. This graphite-epoxy, thin-walled structure is dimensionally sized at each potential mirror operational altitude, that is, for the calculated residual forces, to maintain the specified mirror flatness (in angular measure equal to about $\alpha/10$ where α is the subtense angle of the sun) or, equivalently, the film tension. The torus cross-section and wall thickness is iteratively determined to withstand local buckling and in-plane buckling under the combined mirror tension and torques. The mast wall thickness is determined to withstand local buckling primarily due to the imposed torques. An optimal mast length has been found to be approximately $a/5$, where a is the mirror radius.

At the expense of added constructural complexity, but with resultant decrease in overall mass density, the reflecting film may be divided into smaller panels which are tensioned onto a supportive, in-plane grid which is maintained planar by means of additional guy wires fastened to the mast. At each operational altitude, a panel size can be determined which minimizes the overall mass density σ of this "optimal panel" mirror.

Finally, the means for providing orientational torques has been determined by study of oversite acceleration requirement simulations for a large number of representative mirror passes. These are found to be antisymmetric in time, that is, positive acceleration as the mirror moves from one UVC horizon to the site zenith and a corresponding but negative acceleration as it moves from zenith to the exiting horizon. Thus, the average torques are nearly zero and much smaller than the required peak torques. This characteristic allows advantageous use of reaction wheels, combined with energy storage, which obviates the need for thrusters and their undesirable need for fuel resupply. The masses of these reaction and storage wheel systems, centrally located on the mirror structure, together with a small photovoltaic power supply, and on-board control electronics are added to those of the reflecting film and support structure to evaluate the total mirror mass and the average mass per unit mirror area.

As seen in Figure 6, the areal mass densities can be remarkably small. For smaller mirrors, it approaches that of the thin film reflector itself (ca. 3.5 gm/m^2) for modest altitudes. However, at a given altitude, the penalty for increasing the mirror radius, and therefore, area, is apparent. It is also seen that because of increased mirror loading at lower altitudes, the mirrors must be more robust and thus of higher density.

For a nominal mirror reference in this report (baseline value), it is seen that the density of a single panel mirror, operating at 4146 km, of radius a = 500m and area $\pi/4km^2$, is σ_b=8.76 gm/m^2 or about one-fifth that of newsprint. It is also found that at this altitude, optimum panel mirrors of approximate areas of 30-, 40-, and 50-km^2 have respective values of σ= 10 σ_b, 15 σ_b, and 20 σ_b. These large area mirrors, which would reduce the number of SOLARES satellites to as few as 916 for 4146 km, will be considered later.

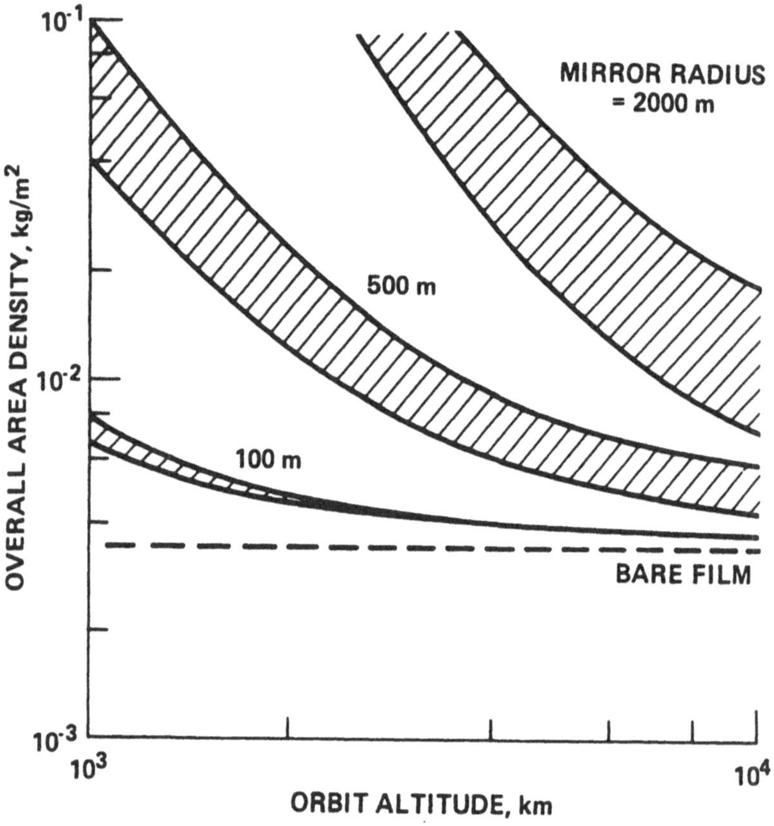

Figure 6. Variation of the Ames mirror overall area density with service altitude and mirror radius as a parameter.

Finally, it should be noted that this mirror design, and subsequent conclusions derived from it, are subject to considerable refinement and improvement resulting from studies which are just beginning in this new class of thin film space structures. One such study,[4] by John Hedgepeth of ASTRO Research Corporation, has recently evolved a point design for a similar single panel 1 km structure, exclusive of orientational controls, at substantially lower areal density than that called for by our design. Significantly, it also includes means for thermal compensation and for mirror packaging. It is believed that such a structure could be launched in a single space shuttle and deployed, rather than constructed,in space. This would allow significant and early testing on the most uncertain element of SOLARES - the space component.

Space Transport and Manufacturing

Assessment of the required mirror area, together with our current model mirror densities, allows the determination of the total mirror system mass as a function of operational altitude. As seen in Figure 7, the system mass exhibits a broad minimum in the orbit range of 2,000-4,000 km. Although these masses are large, it is notable that the important measure of a space energy system--mass per unit of electric power produced on the ground -- will be found to be astonishingly low, typically 0.2-0.3 kg/kW for 4146 km equatorial and inclined orbits, respectively.

The space transportation system requirements are basically determined by the mass of the mirror system to be placed in orbit. A number of studies have shown that to install systems in the mass range needed for SOLARES, it is an economic necessity to use a new generation of launch vehicles of the heavy lift class (HLLV). Such completely reusable "space freighters" are predicted to be capable of carrying 400 metric-ton payloads to low earth orbit for as little as $10/kg in the 1990's, in contrast to the expected space shuttle charge of about $1,000/kg.

Since the launch costs are both mass and rate dependent, the most difficult task in assessing the ultimate cost of the SOLARES space system, or ultimately the cost per mirror, has been in this

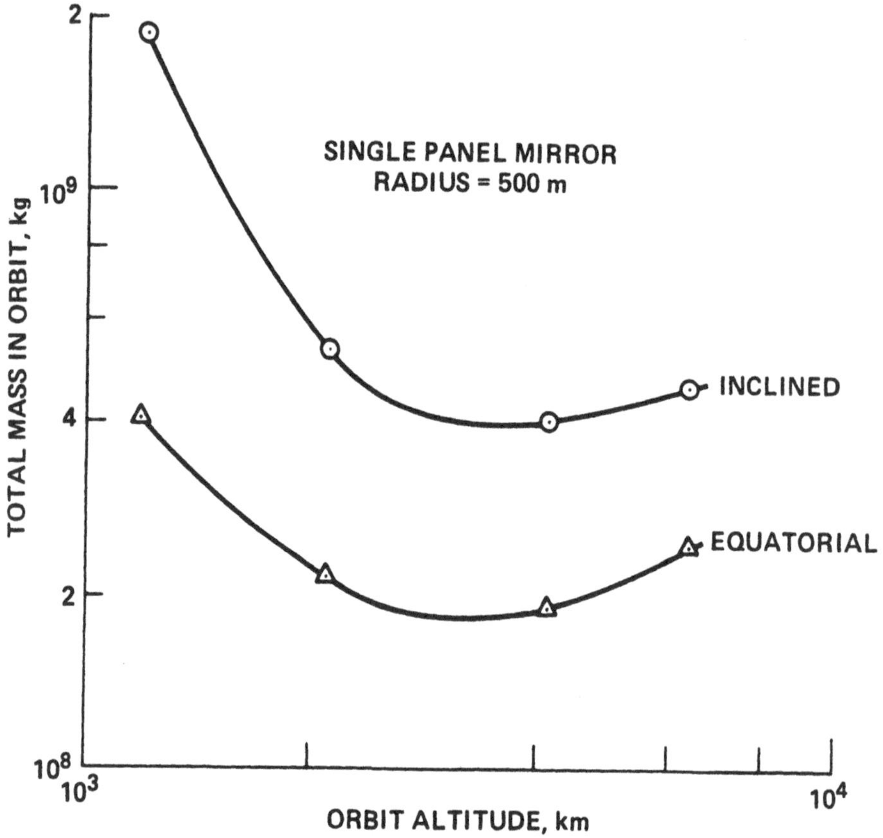

Figure 7. Total mass of the mirror system as a function of service altitude for the inclined and equatorial orbits.

area. Fortunately, extensive studies have been made[5] on a system, the Satellite Power System (SPS) with similar requirements. Our analysis of their HLLV shows that 370 metric-ton payloads could be delivered to 800 km altitude. For a 3×10^8 kg SOLARES system, this would require approximately 811 launches, or about 150 launches per year for a 6-year implementation period. Because this is lower than the 365 launches/year estimated for the SPS, we have accordingly assessed a higher launch cost of $59/kg to SOLARES in contrast to the SPS value of $39/kg.

As shown in Figure 8, three other primary transport needs will be required: (1) a personnel launch vehicle (PLV) to move construction, support, and maintenance crews to and from low earth orbit (LEO), (2) solar sailing from LEO to mirror operational orbit and (3) an inter-orbital transfer vehicle (ITV) to move personnel from LEO to the mirror operational orbits for maintenance and repair operations. The PLV would likely be a derivative of the Space Shuttle, sized to transport a crew of 75. The figure bases construction crew needs on the assumption that the mirrors are largely space manufactured. If it is possible to simply deploy the mirrors, then space crew requirements and the number of PLV's would be greatly reduced.

A key element to space cost reduction appears to be the possible use of solar radiation pressure to provide transport ("solar sailing"[6]) of the mirrors from LEO to their operational orbit. Although this may be faulted in that it is a slow process, typically requiring about 65 days to reach 4,000 km, this is a small "offset" in a system requiring 6 years to construct. Most importantly, it avoids the development of an inter-orbit cargo transport system and the additional orbital burden of it plus its requisite fuel which would have to be carried by the HLLV. Note that the mirrors will be placed into a collision-avoidance configuration, viz., spaced within a given circular orbit and all such orbits will be of slightly different radius. This will avoid the possibility of mirror collision at the "cross-over" points which were indicated in Figure 3. It should also be noted that the solar sailing option should allow a convenient means to produce system expansion by mirror orbit raising, as discussed earlier.

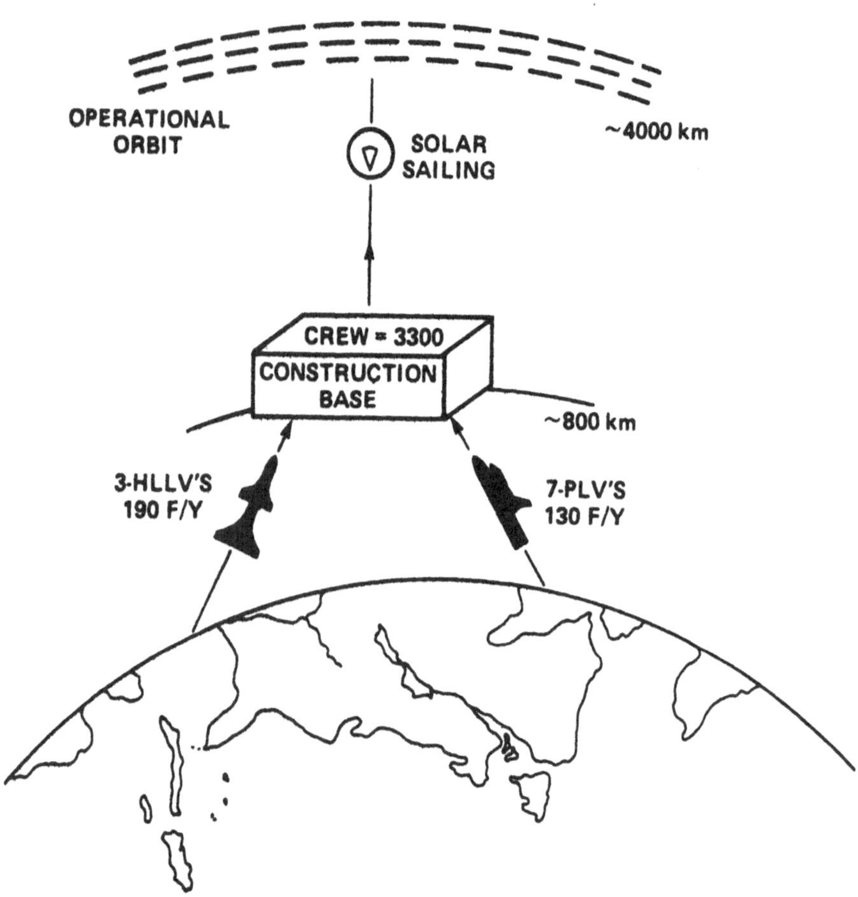

Figure 8. Schematic illustrating the three major transportation elements: HLLV, PLV, and solar sailing and the low Earth orbit construction base which may be required for in-space mirror fabrication.

The Terrestrial Conversion System

A unique property of the SOLARES system is its potential to deliver energy in a form suitable to many direct uses and with the economic advantage associated with continuity and high average energy density, as discussed earlier. Furthermore, as illustrated in Figure 9, it can do this simultaneously, namely, those mirrors over industrial nations can serve their needs, such as electric and fuel production, while at the same time those mirrors over agrarian nations could satisfy such needs as desalination and pumping of irrigation water, fertilizer production, etc. Additionally, in times of emergency, a small fraction of the mirrors could be used to provide large area illumination, frost prevention or snow cover removal.

Although these many uses are potentially important and exciting, limits on the present study have not allowed a true assessment of the cost/benefits of this general SOLARES application. It is clear, as illustrated in Figure 10, that a SOLARES conversion farm will consist of a "mix" of solar conversion technologies which optimize the economics of this most costly element of SOLARES while minimizing the negative environmental effects. Within the $h\alpha$-circle, the highest unit area cost converters, such as solar cells or for special orbits, solar concentrators and converters, will be located. An extensive array of solar-intense biomass growth facilities, desalinators, fuel production facilities, and industries requiring process heat will undoubtedly be placed in the lower intensity outer region of illumination. With controlled sensible heat release, windmills operating with the high efficiency possible with unidirectional and uni-velocity winds, will provide economic means to pump fluids. Finally, the output electricity, fuel, heat, biomass and other products will be linked to the users by a combination of transportation means, such as the supertankers shown.

It is thus clear that the SOLARES site will be a testing, and hopefully, proving ground for the many developments in solar conversion technology. Research dollars spent on this element of SOLARES will notably benefit not only centralized power needs, so necessary for industry, but also will

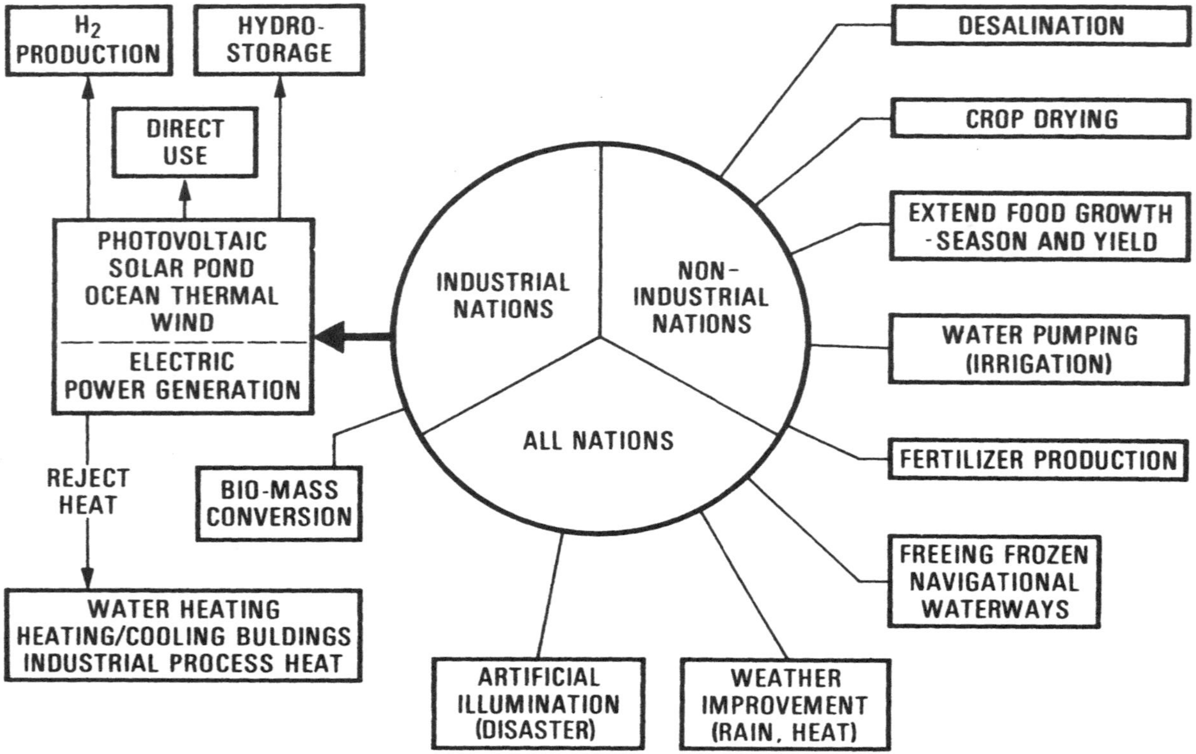

Figure 9. Multiple and simultaneous use possibilities for the solar energy provided by SOLARES.

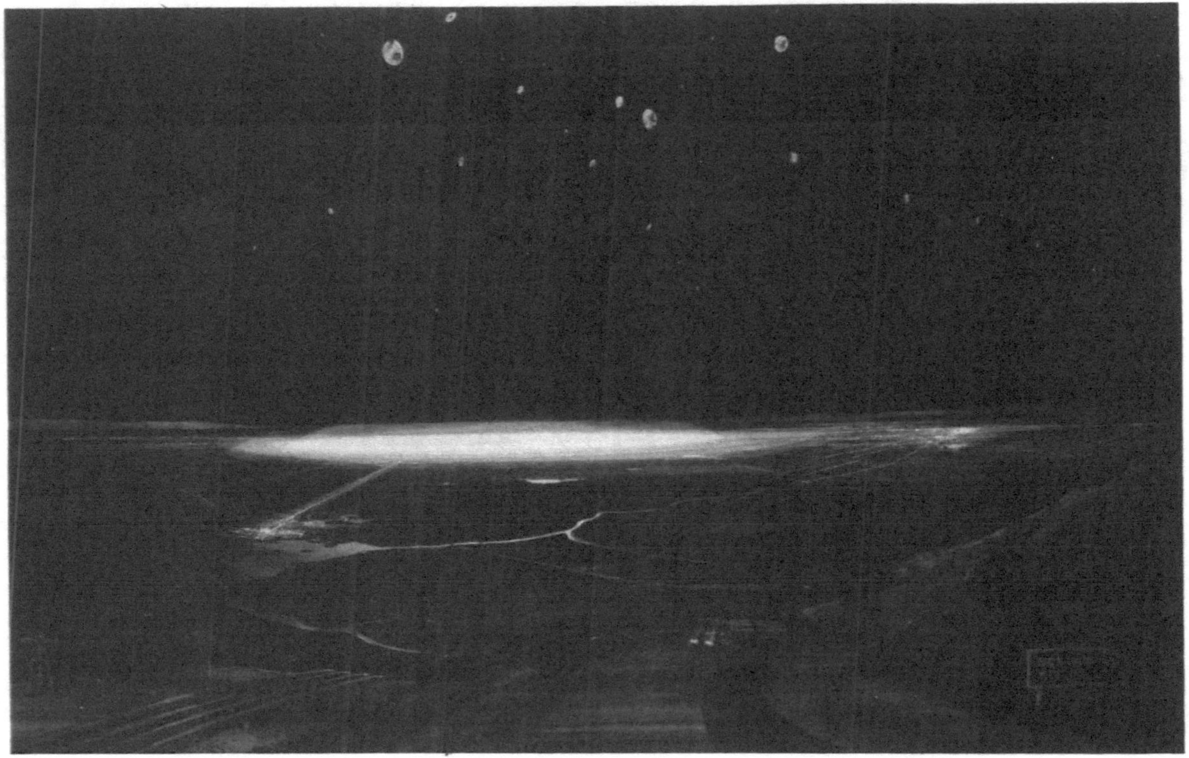

Figure 10. Artist's concept of the multiple conversion "mix" and energy use at a SOLARES conversion site. Energy and product transport is also shown.

produce many solar conversion technical advances and cost reductions (because of the scale of a SOLARES system) which will benefit the small, decentralized use of solar energy.

The SOLARES Baseline Conversion System

Without the benefit of knowing the optimum conversion "mix" for SOLARES, a somewhat "worst case" baseline conversion system has been assessed. This uses horizontal "flat plate" solar cell conversion; heat dispersal (rather than economic use of the "reject" heat) by once-through cooling, electric storage to provide baseload power, and other elements such as power conditioning which constitute a solar photovoltaic conversion system.[7] Conservative (no technology "breakthrough") costing for the 1985-1990 time frame has been assumed. Thus, solar cells are at $500/peak kW, in agreement with the D.O.E. 1986 Goal and near term (battery) electric storage at $36/kWh with 75% depth of discharge giving an effective cost of $48 kWh, etc., are assumed. These, together with the insolation profiles, determine the baseline system costs per unit area of terrestrial farm for each orbital altitude. As seen in the 4146 km summary shown in Table 2, solar cell costs predominate; storage costs are of more significance for the lower, more highly eclipsed, orbits. Also shown are the assumed efficiencies and loss factors which again, with one exception, are those standardly used. The exception is a plant capacity factor, or fractional time of delivery of rated plant output to total available time. This is assumed to be CF= 0.95 in contrast to the standard power industry baseload system value of 0.70. The latter number arises from planned outages (ca. 10%), unscheduled outages (ca. 10%) and system reserve capacity which exceeds routine demand (ca. 10%). Because of the scale of a SOLARES site, however, it is believed that a highly modularized system will be necessary. Fortunately, this is consistent with a photovoltaic power plant but, of course, is not the case for conventional power plants. Routine, automated maintenance, and the storage capacity reserve should allow the SOLARES CF assumed.

An additional, important CF consideration is cloud obscuration. Proper site location usually reduces this to less than 20% for most studies of photovoltaic conversion farms. However, it is our current belief that a controlled, correct sensible heat release from a SOLARES farm will possibly prevent cloud impingement or formation over the site. Gemini satellite photographs dramatically show such a naturally-occurring process at work for Guadelupe Island, located off the coast of Baja California, Mexico. Clouds moving toward the island from the Pacific Ocean are seen to be deflected around the plume of solar-heated air rising from the island. This eliminates obscuration for a site of roughly the same dimensions as a SOLARES site. An important technology challenge to future SOLARES study will be the large scale modeling and computation necessary to understand and make use of these and other possible weather modifications associated with SOLARES.

The Fuel Production Alternative

Recent study has begun to examine the efficacy of fuel production to complement the baselined electric-only system discussed above. Such SOLARES energy output is attractive for many important reasons. Since at many conversion sites electric output would greatly exceed those traditionally connected to the grid, interfacing problems would be reduced if some SOLARES energy were put into chemical form and transported by other means. Secondly, the choice of site location is greatly expanded; equatorial sites, consistent with the lowest cost mirror system and the possibility of high solar concentration and resultant temperature, would be allowed since their energy could be supertanked in the form of methanol or ammonia to points of usage. The environmental impact of SOLARES may also be lessened in this way. Fourth, it must be appreciated that fuels represent an even larger need, by approximately a factor of 4, than electricity for even the industrialized nations. SOLARES might answer this need while avoiding balance of payments deficits and conserving our fossil resources. Finally, such artificial fuel production may decrease the CO_2 buildup in our atmosphere by using atmospheric CO_2 as a chemical feedstock in the synthesis process.

The most desirable means for SOLARES fuel production would be by direct, photo-stimulated reactions. This research area has expanded greatly

Table 2. Elements of the baselined photovoltaic conversion system.

				4146 km orbit case	
	Sizing	Base cost	Efficiency, %	Cost/area, $\$/m^2$	Fraction, %
Solar cells (DOE 1986 goal cells)	$h\alpha$-circle	$500/pk kW	15	75	65
Storage (Battery)	Yearly peak to meet demand curve	$35.7/kWeh	75	16.67	14
Power conditioning	Yearly peak of demand curve	$50/kWe	85	7.80	7
Heat dispersal (once-through)	Adequate to meet yearly peak at $\Delta T = 20°C$	CWP: $\$1/m^2$ Pumps: $\$0.04/m^2$ H. Ex. $\$4.96/m^2$	--	6.29	5
Structure, site preparation and wiring	$h\alpha$-circle	$\$10/m^2$	--	10	9
		Total cost/array area		$115.76	100

Table 2. (continued)

Efficiencies:	Photovoltaic array	$\eta_{PVA} = \eta_{PV}\eta_A\eta_W\eta_T = (0.15)(0.95)(0.90)(1.0) = 0.1283$
	Overall System	$\eta_{SYS} = \eta_{PC}\eta_{PVA} = (0.85)(0.1283) = 0.1091$

Capacity factor: CF = 0.95
Total site size: *hα*-circle + buffer = 30 km (19 mile) radius
Number of sites possible: 6
Total site cost: $135.5 B
Yearly energy output/site = 1.18×10^{12} kWeh
Average power output/site = 135 GWe
Cost/rated output power[a] = $1304/kWe

[a]Includes R&D, facilities, hardware, 15% contingencies, and interest during construction at 8%/annum.

recently and will possibly evolve means, coupled with the economic advantage of SOLARES' continuous radiation, to be of significant importance. A second conversion approach would be thermal. Finally, the more expensive method, but one of present engineering and economic predictability, is that of solar-to-electric conversion followed by the use of electric power for chemical production.

The production of two chemicals: ammonia and methanol by the latter method have been examined. Each uses circa 1985 predicted achievement in high pressure electrolysis of water to obtain the hydrogen feedstock necessary for synthesis. Although it has been frequently argued that hydrogen could be directly transported for fuel usage, the expense and safety problems associated with cryogenic supertanker transport are avoided with methanol, and, to a lesser extent, ammonia. A high temperature/pressure catalyzed reaction of hydrogen with atmospheric nitrogen produces ammonia. Dugger *et al.*,[8] have evaluated this process and its economics. Catalytic combination of the hydrogen with CO_2, obtained from the atmosphere or by steam-stripping from the carbonates in sea water, produces methanol as examined by Steinberg and Dang.[9,10]

Significantly, the additional SOLARES plant capital investment necessary for such fuel synthesis is only 5-15% of the baseline electric production system. This results in quite acceptable costs of about $130/ton for ammonia (currently costing about $150/ton using cheap natural gas as feedstock) and $17/bbl for methanol (currently costing about $40/bbl).

Significant and Renewable Energy -- The SOLARES Promise

An important measure of the worthiness of pursuing a new energy alternative is its potential impact upon future energy need. A second consideration is its ability to do so perpetually; i.e., using a non-depletable or renewable energy resource. The rigorous insolation calculations, together with an assumed terrestrial conversion efficiency, allow the SOLARES impact or promise to be assessed.

A particularly useful way to summarize this impact for various possible orbital altitudes is to consider the following system buildup scenario. Mirrors are continuously added to the system and their altitude raised while four conversion sites simultaneously are continually expanded in radius. At the beginning, the modular sites operate on ambient solar energy to accrue revenues. After 1.6 year, the ground sites have reached the size necessary and the mirrors, at 2109 km altitude, are sufficient to allow operation at an average 1.14 kW/m^2 radiant intensity. Continuing this process, a 4146-km system would be obtained at 6 years. Finally, if desired, further expansion to a 6384-km system would be attained at 15 years.

The impact on world energy needs of such a scenario would be significant. Figure 11 compares the radiant energy/year falling on the four SOLARES sites with the predicted world energy needs.[11] Depending upon the degree with which the SOLARES energy can be converted to use, by the proper "mix" of solar conversion and "reject" heat technologies that are evolved, the potential for such a system to supply a large fraction of world energy can be appreciated. Two U.S.-owned sites, if we allow islands or other ocean bases, may be possible. Additionally, realizing that the U.S. uses about one-fourth of the present world consumption, this study allows us to appreciate the potential impact of the total four-site output of this SOLARES system on future global needs.

If we make the conservative baseline case assumption, however, of only converting the energy to electrical output and with overall system conversion efficiency of 11%, then the effect of <u>one</u> site on the predicted U.S. consumption[12] is seen in Figure 12. With advanced solar cells projected to be available in the 1990's[13], the higher impact shown would result.

Economic Feasibility -- The SOLARES Challenge and Promise

The economic challenge to the SOLARES concept is, as discussed earlier, to provide sufficient economic improvement to solar energy conversion so that it attains competitiveness with alternate fossil and nuclear generation. In order to

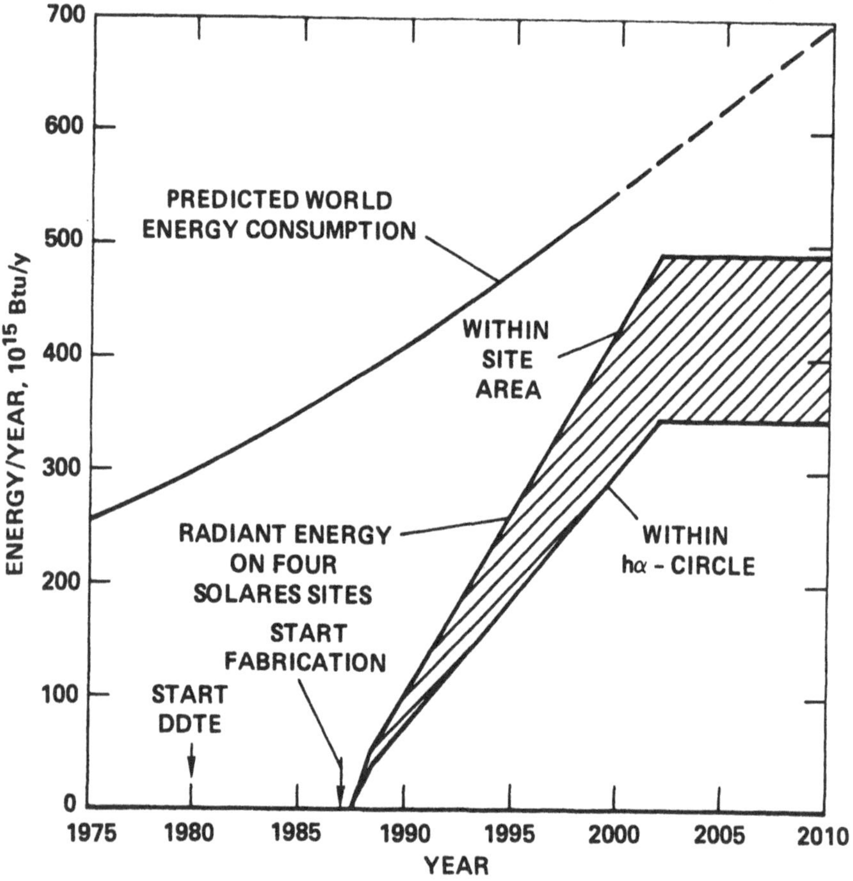

Figure 11. A comparison of the predicted world energy consumption with the radiant energy which would be supplied by a SOLARES system expanded as described in the text.

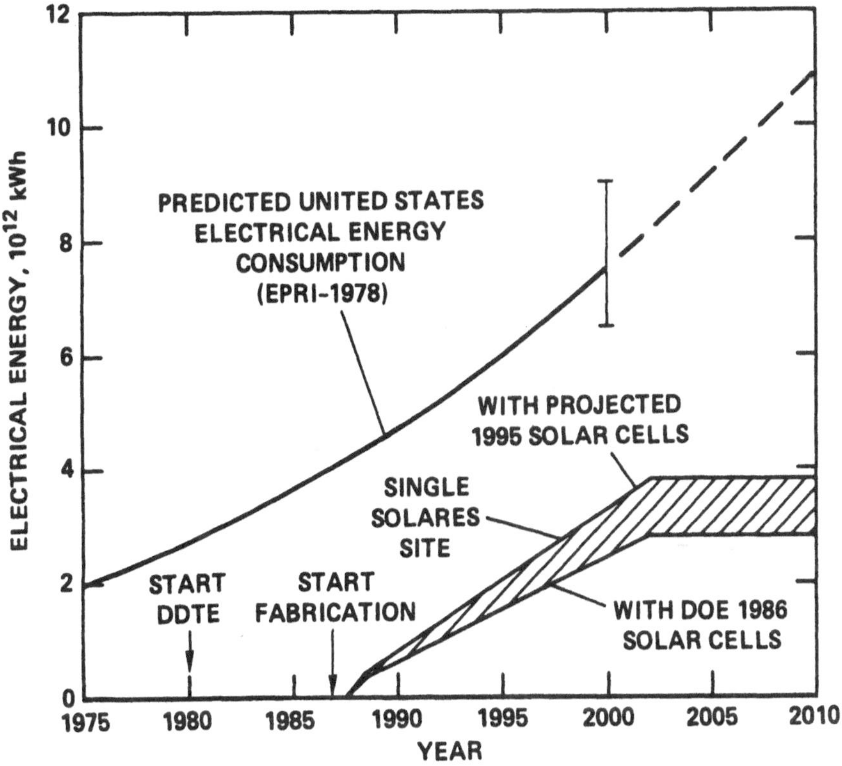

Figure 12. Potential impact of the electric output from a single SOLARES site on predicted U.S. needs. (Uniform growth scenario, as described in the text.)

Table 3. Selected baseload generation candidate alternatives - mid 1990's. (In 1977 $)

Plant type	Direct coal burning[g]	Nuclear[h]	Combined cycle coal liquid[i]	Fluid bed combustion[j]	Gas boiler – advanced gasifier[l]	Combustion turbine – advanced gasifier[l]	Advanced fuel cell advanced gasifier[l]
Capital cost ($/kW)[a]	700	830	345	800[k]	795[k]	730[m]	730[k]
Capacity factor (%)	70	70	70	70	70	70	70
Average heat rate (Btu/kWh)	10200	10400	8100	9800	12200	9100	9100
1995 O&M costs							
Fixed ($/kW-y)	2.53	2.86	1.20	10.99	10.93	14.39	13.59
Variable (mills/kWh)[b]	4.16	0.72	1.27	5.26	1.86	2.11	2.02
1995 fuel cost ($10⁶/Btu)	1.11	.777	3.75	1.11	1.11	1.11	1.11
Levelized capital cost (mills/kWh)[c]	17.1	20.3	8.4	16.5	19.5	17.9	17.9
Levelized O&M cost (mills/kWh)[d]	8.6	2.2	2.8	13.3	6.9	8.4	8.0
Levelized fuel cost (mills/kWh)[e]	23.7	19.9	59.8	22.7	28.3	21.1	21.1
Levelized busbar energy cost (1977 mills/kWh)[f]	49.4	42.4	71.0	52.6	54.6	47.4	47.0

Data source: EPRI Technical Assessment Guide, EPRI-PS-866-SR, June 1978

[a] Includes all direct costs, 15% for contingencies, interest during construction at 8%/annum.
[b] Includes consumables.
[c] 15% fixed charge rate, 30 yr at 6% annual inflation.
[d] 30 yr at 6% annual inflation.
[e] Escalation rate as specified in Technical Assessment Guide.
[f] 15% fixed charge rate.
[g] Plant has flue gas desulfurization.
[h] 0.2% tails, no fuel reprocessing.
[i] Plant burns coal-derived oil.
[j] Plant burns coal.
[k] Technology design/cost *goal.*
[l] Plants operate on synthetic gas produced on site from coal.
[m] Preliminary process design cost.

examine this possibility, it will be necessary to establish the predicted energy costs for these alternatives in the 1990's -- the possible SOLARES implementation period. Assumption of a baseline SOLARES system and current best-estimate, and frequently non-optimized, costing for it will then allow a similar "baseline" energy cost for SOLARES. Comparison between these will then allow a means to examine the sensitivity of SOLARES' competitiveness to uncertain cost elements, such as space costing, and to certain technology options, such as mirror size.

It is particularly important to examine concepts, such as SOLARES, which have not yet had in-depth systems and economic analyses in this fashion because it minimizes the importance of subjective and tentative costing while maximizing insight into essential economic viability of the concept in comparison with its competition.

Busbar Costs of the Alternatives

The accepted measure of economic comparison between alternative electric power systems which serve the same function, such as baseload output, is the busbar energy cost, levelized over the operating lifetime of the system.[14] The levelized busbar costing methodology, as elucidated by Doane, *et al.*,[15] determines that representative price per unit of energy which, if held constant throughout the life of the system would provide the required revenue, assuming that all cash flow interim requirements or excesses are borrowed or invested at the utility's rate of return. The algorithm used to determine the $\overline{BBEC}$ correctly adjusts all costs, expressed in current dollars, for inflation, and in some required cases such as for fuel costs, for escalation. Proper accounting of the total revenue needs, both capital and interest, during construction are assessed. Unfortunately, space will not permit a full discussion of this well-documented[15] method of levelized busbar energy cost, $\overline{BBEC}$, calculation.

Table 3 summarizes the key elements used, and the calculated values of $\overline{BBEC}$ for the most probable baseload generation alternatives (as listed by EPRI) for commencement of operation in the mid-1990's. All costs are expressed in 1977 dollars.

Table 4. SOLARES baseline system.

Configuration:	SPACE SYSTEM 4146 km inclined orbit, 45,800 km² total mirror area		
	GROUND SYSTEM 6 sites with DOE 1986 goal solar cells @ 15% efficiency, 11% overall system conversion efficiency, *hα*-circle area = 1,168 km² each, 135 GWe each		
Impact:	Total system would produce 3.24 times current U.S. consumption, total area = 84 × 84 km² (52 × 52 mi²)		
Baselined costs (in 1977 dollars):	IMPLEMENTATION SCHEDULE 5 yr development, design, test and evaluation (DDTE) 2 yr manufacturing and transport fleet facilities preparation 6 yr space and ground hardware construction		
	System complete about 1995		
	DIRECT COSTS ESTIMATE ($ B)		
	Facilities	$ 47.30	
	Hardware	885.65	
	Total direct		$932.95
	INDIRECT COSTS ESTIMATE ($ B)		
	15% contingency on direct costs	$139.94	
	Design, development, test and evaluation	43.80	
	Interest[a]		
	Facilities	23.58	
	Hardware	101.26	
	DDTE	41.01	
	Total indirect		349.59
	TOTAL COST		$1282.54
	INDIRECT COST FACTOR	1.38	
	INSTALLED COST PER RATED OUTPUT ($/kWe)[b]		1508
	CAPACITY FACTOR (%)		95
	1995 O&M COSTS		
	Fixed ($/kW-y)		3
	Variable (mills/kWh)		2
	LEVELIZED CAPITAL COST (mills/kWh)[c]		27.2
	LEVELIZED O&M COST (mills/kWh)[d]		4.5
	LEVELIZED BUSBAR ENERGY COST (mills/kWh)[e]		31.6

Table 4. (continued)

COMPARISON BASELOAD POWER SYSTEMS (CIRCA 1995):	
Conventional coal/nuclear mix[f]	
Levelized busbar energy cost (mills/kWh)[e]	45
Ambient sunlight photovoltaic[f,g]	
Levelized busbar energy cost (mills/kWh)	115

[a]4% first yr, 8%/annum until positive cash flow after year 11.

[b]Includes all direct costs, 15% contingency, interest during implementation at 8%/annum.

[c]15% fixed charge rate, 30 yr at 6% annual inflation.

[d]30 yr at 6% annual inflation.

[e]15% fixed charge rate.

[f]See text; these do not include their historically extensive R&D costs which are included, however, in SOLARES costing.

[g]Uses same terrestrial costing algorithm as SOLARES which results in indirect cost factor of 1.37.

Input data for the plants was obtained from the EPRI Technical Assessment Guide of June 1978 [14] and from DeMeo, et al.[7] It should be noted that the EPRI capital cost per installed kW estimates shown may appear low in comparison with recent press accounts of certain potential installations, such as the Sun Desert nuclear facility in California, estimated at $1700/kW, the coal power plant at Four Corners at $1120/kW, and a nuclear plant in the Philippine Islands at $1700/kW. However, we will use the EPRI costs as the representative "standards of the industry," being mindful however, that environmental and social factors relegating against the burning of coal or the use of nuclear energy may ultimately increase these values.

It is seen that the $\overline{\text{BBEC}}$ values for coal, synthetic coal-derived oil, and nuclear generation range between 42-71 (1977) mills/kWh. We will use a value of 45 mills/kWh, which is representative of a mix of coal and nuclear generation, as that value which SOLARES must attain to achieve economic competitiveness with these alternatives.

To begin this sensitivity analysis, the $\overline{\text{BBEC}}$ for the baseline SOLARES system is assessed. The 4146 km, inclined orbit case which services six sites, three each spaced along 30°N and 30°S latitude, is chosen as the baseline system. The principal cost elements, as seen in Table 4, are the R&D, Facilities (including the space construction facility for assumed mirror fabrication rather than deployment), and Space and Terrestrial Hardware (mirrors and solar photovoltaic conversion farms). As discussed previously, space-related costs were derived mainly from related studies for the SPS,[5] the Outlook for Space,[16] and studies which analyze space manufacturing.[17] The baseline mirror has $\pi/4$ km^2 area and mass density $\sigma_b = 8.76$ gm/m^2. The unit installed cost for this highly replicative structure, which would be produced in large numbers $(45{,}800/(\pi/4) \simeq 58{,}000$ needed) has been assessed at $2M. This represents a unit mass cost of $291/kg. The baseline photovoltaic solar conversion costs, discussed earlier and presented in Table 2, are known with greater certainty. For an $h\alpha$-circle area of 1,168 km^2 (451 mi^2), a unit site cost of $135 B results. Each site has a power rating of 135GWe. Finally, the recent history of power plant construction has shown numerous cost overruns. To allow for this, a contingency allowance of 15% has been included. Of course, design, development, test and evaluation has also been included as an indirect cost. Another important indirect cost, interest on capital during implementation, greatly increases the above. For the rate of expenditure shown in Table 4, the total, overall indirect cost factor then becomes 1.38. This results in a baseline system SOLARES installed cost per rated power output of $1508/kWe.

The resultant baseline system SOLARES $\overline{BBEC}$ is seen to be 31.6 (1977) mills/kWh, as compared with the alternative mean value of 45 mills/kWh. Another comparison of interest is that of the busbar cost for a normal solar farm, operating on sunlight alone, but with sufficient storage to provide baseload output. If costing identical with that used for the SOLARES ground sites is used, a $\overline{BBEC}$ of 115 (1977) mills/kWh results. The clear promise of the SOLARES concept is seen: a solar-electric energy cost reduction of a factor of about 3.6.

Sensitivities to Assumed SOLARES Costs

It is especially informative to examine the allowable range of certain key SOLARES costs which are consistent with the competitive $\overline{BBEC}$ of 45 mills/kWh. Because of the relative uncertainty of SOLARES space costs, and the overriding importance of the highest net cost component of SOLARES -- the solar cells -- these parameters are examined first, as shown in Figure 13. The solid curves indicate allowable solar cell costs per peak kW and efficiencies for various increases in the assumed baseline space cost. For reference, the DOE 1986 goal of $500/pk kW is shown as a band since this goal does not specify the cell efficiency. Since it is generally accepted to be for silicon cells, a conservative AMI value would be 15%. This, of course, is the cell assumed in the baseline SOLARES costing. As seen, a SOLARES <u>space cost increase by a factor of 4 over that baselined</u> (such as due to increased transportation costs, increased mirror density, etc.) <u>could be tolerated and the SOLARES $\overline{BBEC}$ would remain competitive to the fossil/nuclear alternative</u>.

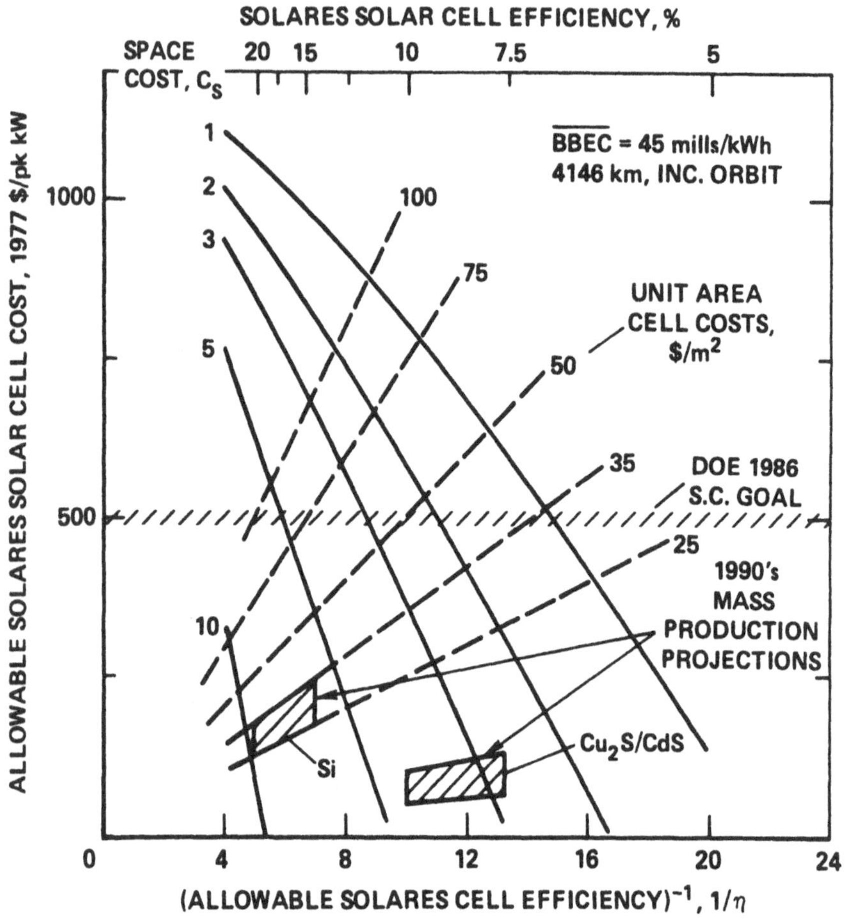

Figure 13. Allowable solar cell cost per peak kW and efficiencies for baselined space cost, C_S and greater, for a 1995 competitive levelized busbar energy cost of 45 (1977) mills/kWh.

As is seen, this is also approximately the case for the projections[7] which have been made on lower efficiency, spray-deposited cells such as Cu_2S/CdS which would appear desirable for SOLARES use. Finally, recent study by Woodcock and others[13] on the cost reductions which should obtain for cell high-production rates (somewhat <u>lower</u> than those needed for SOLARES) and taking into account the base costs associated with material and energy requirements for silicon cells, predicts circa 1990 costs of $\$25-35/m^2$. As seen, these would allow a space cost increase of about 10 and still provide a BBEC of 45 mills/kWh.

Such insensitivity of SOLARES energy cost to the most uncertain cost element is, of course, particularly comforting. A second, and similar analysis examines its sensitivity to the mirror areal density. This parameter, which, of course, is related to the material and transport costs of the mirrors, is important to estimation of allowable unit mirror area and, hence, the total number of mirrors needed in the space system. As discussed earlier, with our present Mark I design, approximately 50 km^2 mirrors would require $\sigma \cong 20\,\sigma_b$. As seen in Figure 14, such would be allowed with the advanced cells predicted by Woodcock.[13] This would reduce the baseline system mirror number to approximately 916 and greatly simplify the logistics of the system. The more conservative cell goals are seen to allow $10\sigma_b$, or mirrors of about 30 km^2. Advanced mirror design will hopefully even improve these numbers.

These sensitivity studies show that large errors in space costing, or alternatively, large reductions in the number of mirrors necessary, are allowable with SOLARES while yet maintaining a competitive economic posture with alternative circa 1990 fossil and nuclear power sources. They help to lend general credence to the promise of SOLARES

<u>Social/Political Acceptability-- The SOLARES Challenge</u>

In addition to the three criteria previously discussed, the ultimate acceptance of a new energy option, and indeed, the continuation of old ones, is closely tied to the social/political desirability of the system. Are the inevitable negative

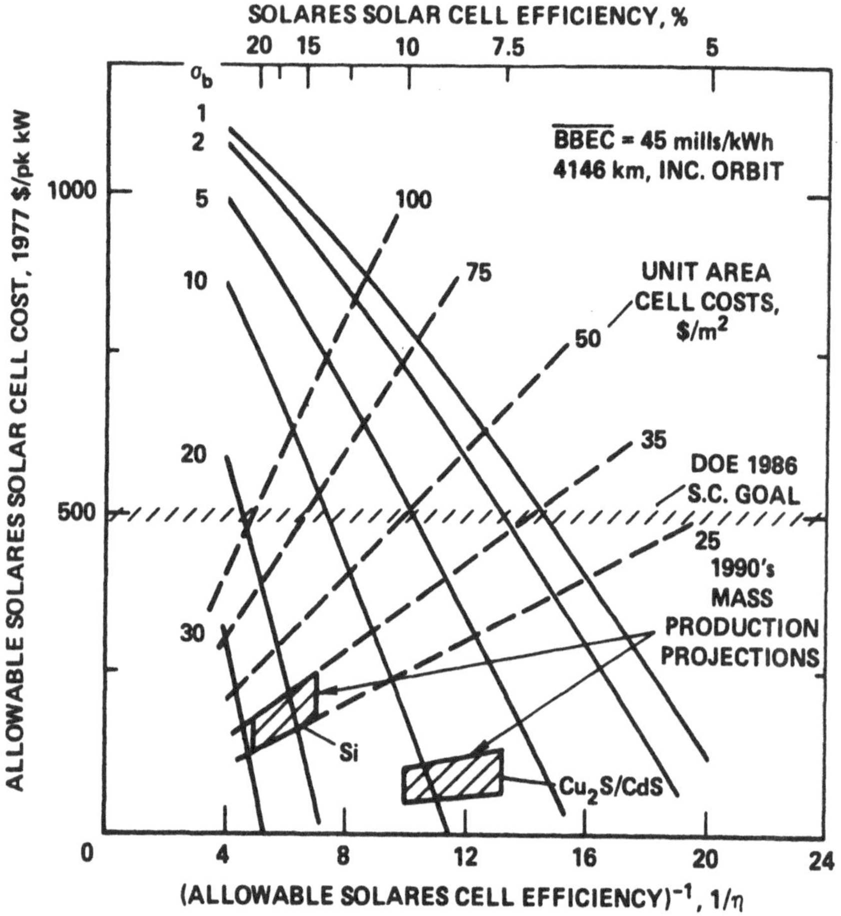

Figure 14. Allowable solar cell cost per peak kW and efficiencies for baselined areal mirror density, σ_b, and greater, for a 1995 competitive levelized energy cost of 45 (1977) mills/kWh.

environmental impacts tolerable and sufficiently outweighed by the positive impacts? Is the economic cost -- in dollars -- and the social cost -- in committed man years -- acceptable? Will the system promote domestic and world stability or will it disrupt it and, in fact, thus be a fragile target for terrorist or military disruption? Finally, and of course ultimately, is the system really perceived as necessary, and if so, is it the best choice between other options?

Complete answers to these questions in relation to SOLARES are, of course, beyond the scope of this paper. However, certain facts are clear and will be briefly discussed.

Environmental Impact

As with any system of the magnitude of SOLARES capable of generating a large fraction of the world's energy needs, environmental impact is inevitable. Those which have been identified are seen in Table 5, listed in positive, negative, and uncertain categories. Clearly, a positive impact will be to allow conservation of non-renewable fossil fuels which are currently used for electric power generation. In addition, SOLARES power may also be used for other applications such as transportation, for which fossil fuels are now the only economical option. Additional conservation would accrue if the solar farm reject heat were used for such purposes as space heating, pumping of fluids, water desalination, industrial process heat and crop drying. Significantly, in addition to fossil fuel conservation, SOLARES would also eliminate the chemical and thermal pollution due to the extraction, transporting, and burning of these fuels or similar problems existing with nuclear fuels.

Limited investigation of the potential photochemical effect of the added SOLARES solar radiation in the atmosphere shows very little net disturbance; in fact, a slight positive or negative change in ozone production, depending upon the uv reflectance cutoff of the mirrors, was found by numerical simulation.

The question of weather modification must await large scale numerical computation on weather models which contain the important fluid physics

Table 5. Identified SOLARES environmental impacts.

POSITIVE	
• Conserve fossil fuels for non-electrical needs	These decrease chemical pollution due to mining, transporting and burning fossil fuels
• Use electricity for *new* uses now only provided by fossil fuels	
UNCERTAIN	
• Photochemistry	
• Weather modification	
NEGATIVE	
• Land usage	
• Light glint and scattering	
• Space transportation pollution	
• Sensible heat release	

and the scales of the SOLARES system. Unquestionably, the possibility of weather improvement is potentially one of the most important uses for an orbiting mirror system. Simple, but important, applications such as removal of fog from critical transportation areas (airports, harbors, etc.), prevention of killing frost to certain expensive crops, extension of growing seasons, and snow cover removal, appear possible. More subtle changes may be effected by the addition of the SOLARES energy with the proper timing and in the proper global location. These possibilities must await future detailed and very complete study.

The potential negative impacts are also listed in Table 5. Generally, they await full technical assessment and with this, hopefully, technology "fixes." Historically, the land usage associated with extensive solar-electric production has been cited as a fault of this energy option. However, because SOLARES provides a five-fold increase in average energy per unit area over that of normal sunlight, this is not the case here. In fact, SOLARES land usage, including a buffer zone of an extent to include all reflected insolation down to 0.1 of the central intensity, would be a factor of 4.4 less than normal solar farms and would approximately equal, per unit of power generated, that of fossil and nuclear plants. For the baseline system which would exceed present-day world electrical consumption, a total area of only 138 X 138 km^2 (86 X 86 mi^2) would be needed.

Without further study and optimization, the remaining impacts are more difficult to assess. One is the visual effect of light scatter from the mirrors and from the site. The former would be of particular concern to astronomers. The latter would be similar to the general glow seen above a desert city at night, and observable out to about 150 km. Most scattering above the site is "small angle" and originates from particulates below 3 km altitude. Because of the scale of the site, virtually all of the scattered radiation will fall on the site. It should be appreciated that in all scattering situations conceived, either from a single mirror which inadvertently sweeps a reflected sun image across the ground during an orientational maneuver (called "glint") or in the cases mentioned above, the intensity of the light is very small and could produce only a visual effect. For reference, it must be remembered that the full, collective intensity of all the mirrors above the UVC is only that present in the desert at noon. The chemical and noise pollution of the launch vehicles necessary for implementation must be minimized. Finally, as discussed previously, limits for sensible heat release at the site must be established. These will enable the design of a conversion "mix" which will minimize undesirable weather modification by SOLARES.

Social-Economic Impact

Clearly, the implementation of a new energy source which effectively replicates much of what now exists would have profound effects upon the society which produces it. Men, material, and financial resources would be required in a measure which has few parallels in peace-time endeavor. In a recent interesting paper,[18] Vajk, et al., examine the various means for the management and financing of a system (the SPS) of similar magnitude. Reasonable methods, involving combinations of governmental, corporate, and public institutions, do appear possible.

To place the SOLARES financial needs into perspective, we recall that the baseline system cost was about $5.9 B/yr for the five years of research and development, $17.6 B/yr for two years of facilities preparation and $185 B/yr for six years of space and terrestrial hardware production. This sums to $1,173 B or an average yearly value of $90 B(excluding interest). For comparison, this value is less than yearly monies collected into the U.S. social security system; it represents about 20% of the federal budget; it is roughly 5% of the U.S. GNP; it is only three times the estimated annual investment of the U.S. electric power industry needed to just keep up with increases in U.S. demand. With the inclusion of other benefiting countries into the supply of capital, if desired, the cost of a significant shift to a solar energy economy would, therefore, appear reasonable. Alternatively, with sole U.S. financing at a higher burden, the sale of SOLARES insolation would again establish this country as a major exporter of energy. Gross revenues from the baseline system would be $262 B per year at

37 mills/kWh, 15% rate of return. At 45 mills/kWh, they would be nearly $400 B/yr.

The Minimal Cost System

It should be noted that the baseline system does not represent the lowest capital investment SOLARES system. Although a full study of the modified space costing for the hardware of a low altitude (1177 km) equatorial orbit has not yet been completed, it is clear that it will be much smaller than the baseline since it requires only one-seventh as many mirrors. Higher requisite mirror areal density removes some of this advantage, however. The estimated cost (including identical R&D and facilities costs) would be $148 B for mirrors plus one ground site. Each additional 10.4 GWe site would cost $12.6 B; a maximum of ten could be serviced. Such a system would find its primary use in fuel production, as discussed earlier, and its yearly output would be equivalent to about 10^9 bbl of oil. The present U.S. oil import is about 3×10^9 bbl/yr. Thus, even this "Minimal Startup SOLARES System" would significantly help to reduce the approximately $110 M/day contribution of oil to the trade deficit.

Can We Afford "Free" Energy? -- The Interstate Highway System Analogy

As has been shown, the cost of the baselined system is very large, in consonance with its potential energy impact. A crucial concern is the availability of this capital -- the "front end money" -- necessary to implement it and thus allow the 30-yr return on investment which the competitive busbar cost would give. There are many possible ways to reduce this initial investment, such as beginning with the aforementioned minimal system and later expansion. Another is to recognize the potential importance of this solar energy option to the country's needs and to support much of the research and development from general funds. This, of course, has been extensively done in the area of nuclear fission reactor research for the past 35 years, and hence, these tax supported costs do not appear in the unit cost of $830/kW quoted earlier. Since much of the SOLARES R&D would be of multiple benefit, this approach appears justified.

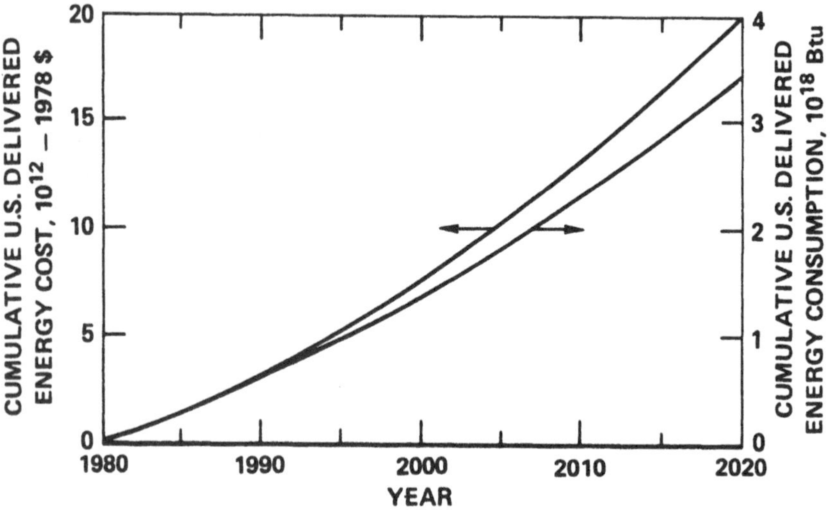

Figure 15. Projected cumulative U.S. consumption and expenditure for delivered energy from the year 1980.

There is an interesting analogy between financing a new energy source like SOLARES and that of the country's interstate highway system. For both military and economic reasons, the taxpayers and their representatives decided to accrue revenues which would result in the future "free" use of these roads. The total cost of the 42,500 mile system will be about $170 B in 1978 dollars; it is presently 92% complete. These revenues have been collected over a 25-year period by taxes on motor fuel (4¢/g), commercial vehicles and parts, tires, oil, etc. -- i.e., they are a "potential user tax." Thus, there is precedent for the accumulation of very large capital sums in an equitable way from those who will benefit by its use.

A similar method could be used for SOLARES, or other new energy source, which would be based upon the concept of an energy replacement tax: those who consume energy (mostly non-renewable fossil-derived) would be assessed a tax per unit of energy used which would pay for the development of the means to provide energy for future generations. Although the full implications to the national economy of such a tax are very complex and have not been assessed, an examination of Figure 15 shows its capital gathering potential. The figure shows the predicted cumulative delivered energy consumption and cost to the U.S. users, as calculated from data given in a recent SRI study.[19] For example, during the 1980-1995 period, 983 x 10^{15} Btu of energy will be consumed at a total cost of 5,270 X 10^{9} 1978 dollars. A 1% surtax would thus give $53 B; 5% giving $264 B. If the period were extended to 2025, that is, the end of a nominal "book life" of a power system beginning operation in 1995, these numbers would be $235 B and $1,175 B, respectively. In comparison, the estimated (baseline) cost of all (both space and terrestrial) research and development, DDTE, and the space components of SOLARES is $84 B, exclusive of interest. Each of the six ground conversion farms would add $158 B, for a total of $948 B.

From these numbers, without benefit of extended economic analysis of the proper mix of investments, spending, interest on borrowed money, etc., which such an energy replacement fund (or "international energyway system" fund, in analogy with the interstate highway system fund) would follow, it is estimated that capital in sufficient quantity could be available to support SOLARES. A 1-3% tax would allow the fund to provide the mirror system and then sell the SOLARES insolation to the entities (public, private, or international) which finance the ground sites; a 5-7% tax might allow two of the sites to be purchased for domestic "free energy" as well.

The "sale" or exportation of the radiant energy supplied by SOLARES is a concept which raises a number of interesting points. Foremost is "what is its value?" An approximate answer can be obtained by comparing the levelized busbar cost of energy which a SOLARES plant could competitively charge, 45 mills/kWh as discussed earlier, with that which would necessarily be charged to simply obtain return of capital ground site investment and pay operations and maintenance. With a unit cost (terrestrial cost only) of $1304/kW, the latter number is 25 mills/kWh. Thus, the value of SOLARES radiation, essentially "fuel" to the plant owner, is 45-25=20 mills/kWh. Charging such a rate would return levelized revenues of about $150 B/yr to the energy trust fund -- a revenue which would allow eventual mirror system expansion. Since at least four sites would probably be foreign-owned, energy export to these would also help alleviate a balance of trade deficit (currently about $15 B/yr).

Again, it must be emphasized that these numbers are only preliminary estimates. However, they indicate one method, related to the replacement of energy used, which appears to be available for accumulating the monies necessary for a new world energy system such as SOLARES.

Conclusion

The SOLARES concept has been examined under the four criteria which any candidate energy system must satisfy. Our assessment of the technical feasibility indicates that although there are many challenges which remain for further detailed study and resolution, no "showstoppers" have been found which would demand other than reasonable extension of current space technology and terrestrial solar conversion methods. Recent, independent investigation by others on low mass density mirrors of the SOLARES genre, which may be deployable and scaled to larger areas with lower mass than our design, is particularly encouraging. It is believed that similar progress will be made in the solar conversion area when the opportunities afforded by the nearly continuous, higher average radiant intensity of SOLARES is fully appreciated. In this way, the economically and environmentally desirable mix of conversion methods, rather than the expensive, electricity-only method of photovoltaic conversion which was used here to expeditiously "baseline" a system, will result. The second criterion examined, significant and renewable energy impact, showed the real promise of SOLARES to supply energy to the world community. If the system were expanded to the size discussed, the radiant energy supplied would exceed present, total world energy use. Our ability to evolve a high conversion efficiency or total-use system at the sites will determine the ultimate utility of this SOLARES promise. Significantly, even with the low efficiency of the baselined system, a single, highly modular site would provide about 40% of the predicted U.S. electrical needs in year 2002. The third criterion examined, economic feasibility, was approached in a manner consistent with our present lack of in-depth systems and economic assessments. A baselined system, with space and terrestrial costs consistent with studies by others, was compared with the circa 1995 candidate baseload power systems listed by EPRI. It was found that with projected 1986 DOE solar cells, and certainly with more advanced cells predicted by others, large overruns in the most uncertain cost element of SOLARES -- the space component -- would not change the competitive posture of SOLARES-derived energy to the fossil/nuclear mix alternative. On the other hand, if the baselined costs could be achieved, greater technical flexibility results -- such as the use of higher mass, very large mirrors which would significantly reduce the number of satellites required. Finally, a brief examination of the fourth criterion, social/political acceptability, showed the many questions yet remaining in this area. The positive impacts of conserving fossil chemicals and thereby avoiding their pollution in power generation or transportation use are believed important.

However, the most significant unanswered question about SOLARES -- that of possible weather modification -- must be resolved by future study. There is hope that this will show how SOLARES, or some variant of it, could provide beneficial weather enhancement. Certainly, the studies will also define the allowable sensible heat release density in the vicinity of the conversion sites and thus prescribe the conversion and dispersal goals which must be achieved. Finally, the question of the ability to finance this bold step into large-scale, solar-derived energy for the world was examined. One particularly attractive means, the assessment of an energy replacement tax, was estimated to be capable of developing the SOLARES space system for a levy of 1-3% on domestic delivered energy. This investment would again allow our nation to regain its position of energy exporter, providing revenues in excess of $100 B/yr. A 5-7% levy would probably be sufficient to buy two sites and hence provide "free energy" for the populace in analogy to their "free" use of the tax-purchased interstate highway system. Putting such low cost energy into the domestic grid would, it appears, not only improve the quality of life directly, but also benefit the economic posture of our commercial entities in the world market place.

References

1. Billman, K.W.; Gilbreath, W.P.; and Bowen, S.W.; Introductory Assessment of Orbiting Reflectors for Terrestrial Power Generation, NASA TM X-73, 230, April 1977.

2. Billman, K.W.; Gilbreath, W.P.; and Bowen, S.W.; Satellite Mirror Systems for Providing Terrestrial Power: System Concept, The Industrialization of Space, vol. 36, part I, Advances in the Astronautical Sciences, AAS, San Diego, CA, April 1978, pp.391-414.

3. Billman, K.W.; Gilbreath, W.P.; and Bowen, S.W.; Orbiting Mirrors for Terrestrial Energy Supply, Radiation Energy Conversion in Space, K.W. Billman, ed., AIAA Progress in Astronautics and Aeronautics Series, vol. 61, July 1978, pp. 61-80.

4. Private communication, 2 November 1978.

5. Piland, R.O.: The Solar Power Satellite Concept Evaluation Program, Radiation Energy Conversion in Space, K.W. Billman, ed., AIAA Progress in Astronautics and Aeronautics Series, vol. 61, July 1978, pp.3-24.

6. Friedman, L., et al.: Solar Sailing, the Concept Made Realistic, Paper 78-82, AIAA 16th Aero-space Sciences Meeting, Huntsville, AL, January 1978.

7. DeMeo, E.A.; and Bos, P.B.: Perspectives on Utility Central Station Photovoltaic Applications, Solar Energy, vol. 21, 1978, pp. 177-192.

8. Dugger, G.L.; Francis, E.J.; and Avery, W.H.; Technical and Economic Feasibility of Ocean Thermal Energy Conversion, Solar Energy, vol. 20, 1978, pp. 259-274.

9. Steinberg, M.; and Dang, V.; Production of Synthetic Methanol From Air and Water Using Controlled Thermonuclear Reactor Power-I. Technology and Energy Requirement, Energy Conversion, vol. 17, 1977, pp. 97-112.

10. Dang, V.; and Steinberg, M.; Production of Synthetic Methanol From Air and Water Using Controlled Thermonuclear Reactor Power --II. Capital Investment and Production Costs, Energy Conversion, vol. 17, 1977, pp.133-140.

11. Energy Facts II, Science Policy Research Division, Congressional Research Service, Library of Congress, U.S. Government Printing Office, August 1975.

12. Williams, L.J.; Boyd, J.W.; and Crow, R.T.; Demand 77, EPRI Special Report, EA-621-SR, vol. 1, March 1978.

13. Woodcock, G.R.: SPS Cost Considerations, J. Energy, vol. 2, July-August 1978, pp.196-202.

14. Technical Assessment Guide, EPRI Technical Assessment Group, EPRI Special Report, PS-866-SR, June 1978.

15. Doane, J.W., et al.: The Cost of Energy from Utility-Owned Solar Electric Systems, JPL Report JPL 5040-29, Jet Propulsion Laboratory, Pasadena, CA, June 1976.

16. Anon.: A Forecast of Space Technology 1980-2000, NASA SP-387, Scientific and Technical Information Office, Washington, DC, 1976.

17. Davis, E.E.; and Miller, K.H.: Construction of a 10 GWe Solar Power Satellite, 13th Intersociety Energy Conversion Engineering Conference Proceedings, SAE, Warrendale, PA, 1978, pp. 189-194.

18. Vajk, J.P.; Stutzke, R.D.; Salkeld, R.; Stine, G.H.; and Klan, M.S.: Satellite Power System (SPS) White Paper on Financial/Management Scenarios, Science Applications, Inc., Pleasanton, CA, 10 October 1978.

19. Fuel and Energy Price Forecasts: Quantities and Long Term Marginal Prices, EPRI Report EA-433, vol. 1, September 1977.

Overview

Robot and automation technology, as outlined by Ewald Heer, are destined to play a significant role in future space missions. Increased reliance on robotics and automation offers the likelihood of enhancements of mission capabilities along with reductions in overall costs. Areas where this technology can be effectively employed include planetary exploration, global service satellites (such as LANDSAT), and space industrialization. An aggressive robot and automation research and development program in such areas as artificial intelligence, machine intelligence and robotics will greatly facilitate the design of a cost-effective and flexible space program.

Ewald Heer

16. Robots and Automation in the Space Program

Introduction

Most of those who are acquainted with the space program expect that the space transportation system including the Space Shuttle and its adjuncts will open up a new era of expanded space activities in scientific exploration and global service undertakings. The first steps towards utilizing the space environment for industrial and commercial ventures will become possible. This can trigger requirements for more advanced transportation systems in the 1990's that will enable and can lead to large-scale projects. By the end of this century, such projects could include a Space Solar Power System for solar energy production, a lunar base for extracting and processing material resources, and a space station for commercial processing and manufacturing. In addition to developing the necessary transportation systems, a major objective for the space program is the development of systems that will enable the cost effective construction and erection of such systems in space.

On examining potential space missions during the next two decades, it is evident that robot and automation technology can be a vital contributor to the cost-effective implementation and operation of both scientific exploration missions (earth orbital and planetary) and public service missions (large communication satellites and large scale systems). In some cases, it can make the mission feasible and in other cases, it can make it economically affordable.

This paper briefly examines the possible role that could be played by robots and automation in the space program. It also explores and conjectures on benefits that can be expected as a result.

Background

Looking back on industrial evolution, there are a number of key steps: division of labor, use of mechanical power, standardized parts, transfer devices, open-loop control, closed-loop control, and computer control. These steps indicate a progression in the amplification of human capabilities. Until the dawn of the industrial revolution, human physical capabilities (muscle power) were amplified through the use of simple machines. Then automation also came into the picture. After a slow beginning, automation has been moving on with ever-increasing pace, helping to accomplish more and more human tasks. By means of limit switches and governors, the machine became able to make simple decisions on its own, such as "go-stop" or "more-less". Some outstanding examples are assembly line transfer devices in automobile manufacturing during the early twentieth century, systems with open-loop mechanical control around World War I, and systems with closed-loop control approximately during World War II.

In the 1940's, the advent of the digital computer also opened up the possibility of performing many so-called low-level (and possibly high-level) intellectual activities automatically. In addition to the long-standing amplification of human physical capabilities, the amplification of the human mind was in the making.

Since the first commercial computers were introduced in the early 1950's, the "electronic brain" has brought widespread changes. It has reached into innumerable crevices of our lives. At last count, some 1800 firms and more than 4 percent of our labor force were dedicated to maintaining and extending automation through computers. Today computer-automated operations include collecting, maintaining, and assimilating enormous amounts of data. Automatons greatly simplify operations and reduce costs. They pay bills and keep track of accounts, inventories, personnel, and pricing. In factories, many routine or dangerous tasks are done by computer-controlled manipulators. Even some engineering designs and analyses are done automatically.

Not counting the hundreds of thousands of microprocessors proliferating in our society, we have today more than 200,000 free-standing computers. If one added up the performance of these machines in terms of their human equivalent, one would find that it would take more than 400 billion people to accomplish the same tasks that these machines perform today. Considering that, conservatively, at most 20 percent of the U.S. population is engaged in relevant work using these computers, we have built into our society a mind-amplifying factor of about four orders of magnitude. Clearly, America depends on automated machines so much that the country would come to a grinding halt without them. The man-machine symbiotic existence is here, and this relationship is moving with ever increasing speed in the direction of more automation.

Along with the fast pace of these developments several branches of science and technology came into being. Probably computer science emerged as the most encompassing discipline on equal par with other engineering professions such as electrical engineering and mechanical engineering. Along came also changes in the definition and use of old familiar terms. For instance, automation in manufacturing is retaining its usual connotation of self-acting and self-regulating machinery performing specialized pre-programmed tasks; however, in the business world, the words automation and computers as the central element in automated systems are only now becoming dominant, and the newly emerging robot devices are general-purpose systems with a great degree of autonomy. By means of a computer "brain" as the central element, a robot is able to sense its environment, plan and decide its actions, and perform mechanical manipulations and data handling tasks sometimes to a degree normally done by humans. The brain, consisting of computer hardware and software, incorporates the robot's intelligence, i.e., machine intelligence. The developmeat of machine intelligence technology relies heavily on advances in the science themes of artificial intelligence such as problem solving, scene analysis, theorem proving and speech understanding.

Robots and Automation in NASA Planning

While mechanical power provides physical amplification and computers provide mind amplification, telecommunication provides space amplification, i.e., amplification of the space accessible to humans. By means of telecommunication, humans can activate and control systems at remote places. They can perform tasks even as far away as the planets. During the 1960's, this became known as teleoperation.[1] Teleoperators are man-machine systems that augment and extend human sensory, manipulative, and cognitive abilities to remote places. In this context, the term robot can then be applied to the remote system of a teleoperator if it has at least some degree of autonomous sensing, decision-making, and/or action capability. The concept of teleoperation has profound significance in the space program. Because of the large distances involved, almost all space missions fall within the teleoperator definition, and because of the resultant communication delay for many missions the remote system requires autonomous capabilities for effective operation.[2] The savings of operations time for deep space missions can become tremendous if the remote system is able to accomplish its tasks with minimum ground support. For example it has been estimated that a Mars roving vehicle would be operative only 4 percent of the time in a so-called move-and-wait mode of operation. With adequate robot technology, it should be operative at least 80 percent of the time.

Because of the central role of computers in many space systems, both robotics and automation have pervaded the space program, at least to some degree, in many guises since its inception. However, these trends did not take sufficient advantage of the technological developments during the last one or two decades. Some have criticized NASA for this, stating that the computer technology used in NASA missions is at least ten years behind the times. While some of the criticisms are not justified, some may be valid. The valid ones can usually be traced back to the generally accepted requirement of flawless performance of NASA missions (often for political reasons) and to long lead times from project conception to implementation. Projects and missions are usually conceived and their objectives defined based on then proven technology rather than on the technology existing at the time of the project design period.

The development and application of robot and automation technology during the first decade of NASA's existence was hardly organized. Johnsen and Corliss[1] in 1967 presented a comprehensive survey of teleoperator and robot technology to that date. The development of teleoperator technology had its beginnings for serious, practical applications in the late 1940's and early 1950's, when it became necessary to control remotely located manipulators in hazardous environments of nuclear laboratories. With few exceptions, this work was done under the aegis of the Atomic Energy Commission. Nevertheless, in 1966, Case Institute of Technology, working under a NASA grant, demonstrated a computer-controlled manipulator that could perform pre-programmed tasks specified by the operator.

Surveyor 3 and Surveyor 7 landed on the moon in 1967 and 1968, respectively. The claw and arm mechanism of the surface sampler gave the spacecraft the capability of a crude robot. Controlled remotely from earth at each step, it dug trenches and performed soil experiments.

The first organized effort to address the questions of teleoperator/robot technology planning for the space program took place in 1970, when NASA conducted an inter-Center study on requirements for the years 1970-1990.[3] The study team recommended to the NASA Administrator a research and development program with focus on all subsystems (manipulators, mobility units, sensors, displays, controls, communications, artificial intelligence, etc.) and system integration.

Some highlights of the robot technology planning, development, and application activities during the past decade have been presented in Ref. 4.

A series of planning studies and workshops was initiated with the Outlook for Space Study in 1974[5], which included a comprehensive forecast

Figure 1. Galileo spacecraft navigates between Jupiter and Galilean satellites in rendering. After sending a probe into the Jovian atmosphere, the robot spacecraft will perform complex maneuvers at various inclinations with repeated close encounters with the satellites.

of space technology for 1980-2000. In a subsequent NASA/OAST Space Theme Workshop[6], the technology forecasts were applied to three broad mission themes: space exploration, global services and space industrialization. Based on the derived requirements for cost-effective space mission operations, Kurzhals[7] identified five new directions for space electronics technology developments: (1) automated operations aimed at substantial reduction of mission support costs; (2) precision pointing and control; (3) efficient data acquisition to permit a large increase in information collection needed for global coverage; (4) real-time data management; and (5) low-cost data distribution. The machine intelligence and automation technologies for data acquisition, data processing, information extraction and decision making emerge here as the major drivers in each area and call for the systematic exploitation of the spectacular developments in electronics during the past two decades. In addition, for certain areas such as automated operations in space, the mechanical technologies directed at materials and objects acquisition, handling, and assembly must also be further developed; robots doing construction work in earth orbit or on the lunar surface will need manipulative and locomotion devices to perform the necessary transport and handling operations.

Future Applications

In space applications, robots may take on many forms. None looks like the popular science fiction conception of a mechanical man. Their appearance follows strictly functional lines, satisfying the requirements of the mission objectives to be accomplished.

Space Exploration

Space exploration robots may be exploring space from earth orbit like orbiting telescopes, or they may be planetary flyby and/or orbiting spacecraft like the Mariner and Pioneer families. They may be stationary landers with or without manipulators like the Surveyor and the Viking spacecraft, or they may be wheeled like the Lunakhod and the proposed Mars rovers. Others may be penetrators, flyers, or balloons and some may bring science samples back to earth (Figs. 1-3). All have in

Figure 2. Mars surface robot will operate for 2 years and travel about 1000 km performing experiments automatically and sending the scientific information back to Earth.

Figure 3. Artist's concept of a Mars surface scientific processing and sample return facility. Airplanes transport samples into the vicinity of the processing station. Tethered small rovers then bring the samples to the station for appropriate analysis and return to Earth.

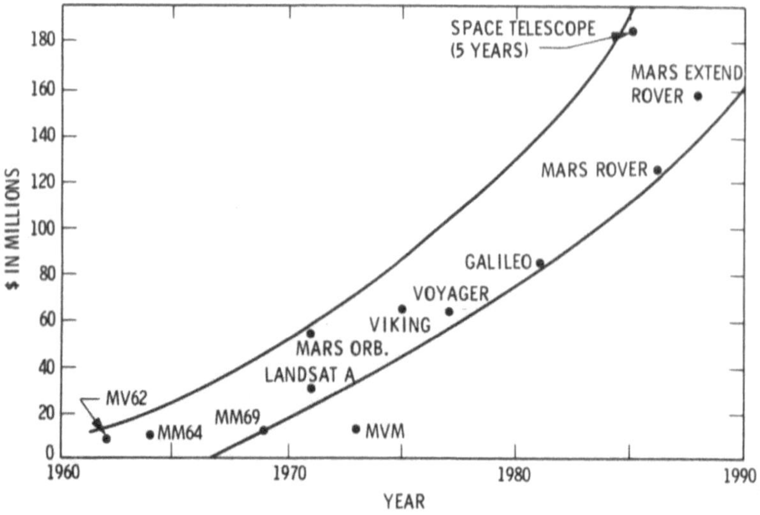

Figure 4. Trend of mission ground operation costs. Increasing mission complexity and duration contribute to the ground operation costs.

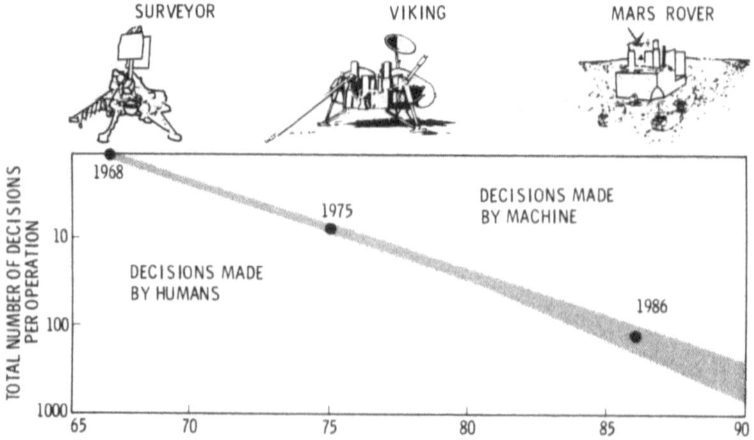

Figure 5. Trend of function (decision) allocation between humans and robots. For the same typical operation, the robot takes over an increasing number of elementary functions, leaving the high-level decisions to humans.

common that they can acquire scientific and engineering data using their sensors, process the data with their computers, plan and make decisions, and send some of the data back to earth. Some robots are, in addition, able to propel themselves safely to different places and to use actuators, manipulators, and tools to acquire samples, prepare them, and perform in situ experiments or bring them back to earth.

Exploration robots are required to send back most of the collected scientific data, unless it becomes repetitive or is meaningless and can be discarded. The unknown space environment accessible to the sensors is translated into a different, still uninterpreted environment, in the form of computer data banks on earth. These data banks are then accessible for scientific investigations long after the space mission is over.

As on-board data acquisition requirements and mission complexity increase, more data is sent to the ground that must be coordinated and handled. If mission durations also increase, the ground-based mission operations will quickly use up a major part of the total mission cost (Fig. 4). With more capable on-board machine intelligence, the robot will be able to perform many of the ground-based operations in space autonomously which would relieve ground activities and would greatly contribute to cost reductions for the same mission operations. Adapting advanced automation techniques to ground-based operations would also lead to significant cost savings. The combined effect of applying advanced automation technology to exploration missions is projected to yield a cost improvement of about two orders of magnitude by the year 2000.

A qualitative measure for the level of machine intelligence is the number of decisions made by the robot versus those made by its human supervisor. For a particular operation like "dig a trench", the total number of elementary decisions is presumably constant at approximately 1000. With increased machine intelligence capabilities, the robot will make a correspondingly increased number of the required decisions (Fig. 5).

Projections into the future lead one to speculate on the possibility of highly autonomous exploration robots in space. Such exploration robots would communicate to earth only when contacted or when a significant event occurs and requires immediate attention on earth. Otherwise, they would collect the data, make appropriate selective decision, achieve it, and store it on board. They would serve as a data bank, and their computers would be remotely operated by accessing and programming them from earth whenever the communication link to the robot spacecraft is open. Scientists would be able to interact with the robot by remote terminal. Indeed, the concept of distributed computer systems, presently under investigation in many places including JPL, could provide to each instrument its own microcomputer, and scientists could communicate with their respective instruments. They could perform special data processing on board and request the data to be communicated to them in the form desired. Alternatively, they could retrieve particular segments of raw data and perform the required manipulations in their own facilities on earth.

Prime elements in the "scientist-deep space robot" link would be large antenna relay stations in geosynchronous orbit. These stations would also provide data handling and archiving services, especially for exploration robots which will not be accessible after a relatively short time, e.g., those leaving the solar system.

Global Services

Global services robots orbit the earth. They differ from exploration robots primarily in the intended use of the collected data. They collect data for public service use on soil conditions, sea states, global crop conditions, weather, geology, disasters, etc. These robots generally acquire and process an immense amount of data. However, only a fraction of the data is information of interest to the ultimate user. At the same time, the user often likes to have the information shortly after it has been obtained by the spacecraft. For instance, the value of weather information is short-lived except for possible historical reasons. The value of information on disasters such as forest fires is of even shorter duration. The demand for high-volume on-board data processing and pertinent automated information extraction is therefore great.

The usual purpose of global services robots is to collect time-dependent data on the earth's environment, whose static properties are well known. The data is used to determine specific patterns or classes of characteristics and translate these into useful information. For instance, for LANDSAT and Seasat A (Fig. 6), the data is presently sent to the ground, where it is processed, reduced, annotated, analyzed, and distributed to the user. This process requires up to three months for a fully processed satellite image and costs several thousand dollars. The image must then be interpreted by the receiver, i.e., the information must still be extracted by the user.

Present developments in artificial intelligence, machine intelligence and robotics suggest that in the future, the ground-based data processing and information extraction functions will be performed on board the robot spacecraft. Only the useful information would be sent to the ground and distributed to the users, while most of the collected data could be discarded immediately. This would require the robot to be able to decide what data must be retained and how it was to be processed to provide the user with the desired information. For instance, the robot's computer brain could have a large number of pattern classification templates stored in its memory or introduced by a user with a particular purpose in mind. These templates would represent the characteristics of objects and/or features of interest. The computer would constantly compare the scanned patterns with those stored in its memory. As soon as something of interest appeared, it would "take a closer look" by zooming in on the pattern and examining it with higher resolution, comparing it to a progressively narrower class of templates until recognition had been established to a sufficient degree of confidence. The robot would then contact the appropriate ground station and report its findings and, if required, provide the user with an annotated printout or image. One can envision that the user would be able to interact with the robot, indeed with his particular instrument, by remote terminal, much the same as with a

Figure 6. Seasat A. The oceanographic satellite's high-data-rate Synthetic Aperture Radar imaging device has provided data on ocean waves, coastal regions, and sea ice.

central computer and, depending on intermediate results, modify subsequent processing.

For space exploration and global services, the ground-based mission operations can become an incredibly complex process. For planetary missions, these mission operations are handled primarily by the Deep Space Network and the Space Flight Operations Facility at JPL.

The latest example of planetary exploration mission, and perhaps the most complex to date, is Viking. At times there were several hundred people involved in data analysis, mission planning, spacecraft monitoring, command sequence generation, data archiving, data distribution, and simulation. While for earlier space missions sequencing had been determined in advance, on Viking this was done adaptively during the mission. The operational system was designed so that major changes in the mission needed to be defined about 16 days before the flight action. Minor changes could be made as late as 12 hours before sending a command. The turnaround time of about 16 days and the number of people involved contributes, of course, to sharply increased operational costs (Fig. 4). The Viking operations costs are for a 3-month mission. The planned Mars surface rover mission is expected to last two years, covering many new sites on the Martian surface. Considering that this mission would be more complex and eight times as long, ground operations would have to be at least ten times as efficient to stay within, or close to, the same relative costs as for Viking.

During the Viking mission, about 75,000 reels of image data tapes were collected and stored in many separate locations. The images are now identifiable only by the time when and the location where they were taken. No indication regarding image information content is provided, and the user will have to scan catalogs of pictures to find what he wants. For such reasons, it is expected that most of the data will not be used again.

The ground operations for earth orbital missions suffer from similar problems [7] as those for planetary missions. The overall data stream is usually much higher here, images are still very costly, and they take up to several months to reach the user.

These considerations lead one to conclude that technology must be developed so that most ground operation activities can be performed as close as possible to the sensors where the data is collected, namely by the robot in space. However, examining the various ground operations in detail, it becomes clear that most of those that must remain on the ground could also be automated with advanced machine intelligence techniques. The expected benefits derived from this would be a cost reduction for ground operations of at least an order of magnitude and up to three orders of magnitude for user-ready image information.

Utilization of Space Systems

Space industrialization requires a broader spectrum of robotics and automation capabilities than those identified for space exploration and global services. The multitude of systems and widely varying activities envisioned in space until the end of this century will require the development of space robot and automation technology on a broad scale. It is here that robot and automation technology will have its greatest economic impact. The systems under consideration range from large antennas and processing and manufacturing stations in earth orbit to lunar bases, to manned space stations, to satellite power systems of up to 100 km^2. These systems are not matched in size by anything on earth. Their construction and subsequent maintenance will require technologies not yet in use for similar operations on earth.

Space processing requires a sophisticated technology. First, it must be developed and perfected, and then it must be transferred into the commercial arena. Basic types of processes presently envisioned are: solidification of melts without convection or sedimentation, processing of molten samples without containers, diffusion in liquids and vapors, and electrophoretic separation of biological substances. It is expected that specialized automated instrumentation will be developed for remote control once the particulars of these processes are worked out and the pressure of commercial requirements becomes noticeable.

Large-area systems such as large space antennas, satellite power systems, and space stations require large-scale and complex construction facilities in space. Relatively small systems, up to 100 m in extent, may be deployable and can be transported into orbit with one Shuttle load. For intermediate systems of several-hundred-meter extension, it becomes practical to shuttle the structural elements into space and assemble them on site (Figs. 7-9).

Very large systems require heavy-lift launch vehicles which will bring bulk material to a construction platform (Fig. 10), where the structural components are manufactured using specialized automated machines (Fig. 11).

The structural elements will be handled by teleoperators or self-acting cranes and manipulators which bring the components into place and join them (Fig. 12). Free-flying robots will transport the structural entities between the Shuttle or the fabrication site and their final destination and connect them. These operations require a sophisticated general-purpose handling capability. In addition to transporting structural elements, the robot must have manipulators to handle them, and work with them and on them. Large structural subsystems must be moved from place to place and attached to each other. This usually requires rendezvous, stationkeeping, and docking operations at several points simultaneously and with high precision -- a problem area still not investigated for zero gravity. Automated (smart) tools would also be required by astronauts to perform special local tasks.

These robot systems could be controlled remotely like teleoperator devices, or they could be under supervisory control with intermittent human operator involvement. Astronauts in space or human operators on earth will need the tools to accomplish the envisioned programs. The technology for in-space assembly and construction will provide the foundation for the development of these space-age tools.

After the system has been constructed, its subsequent operation will require service functions that should be performed by free-flying robots or

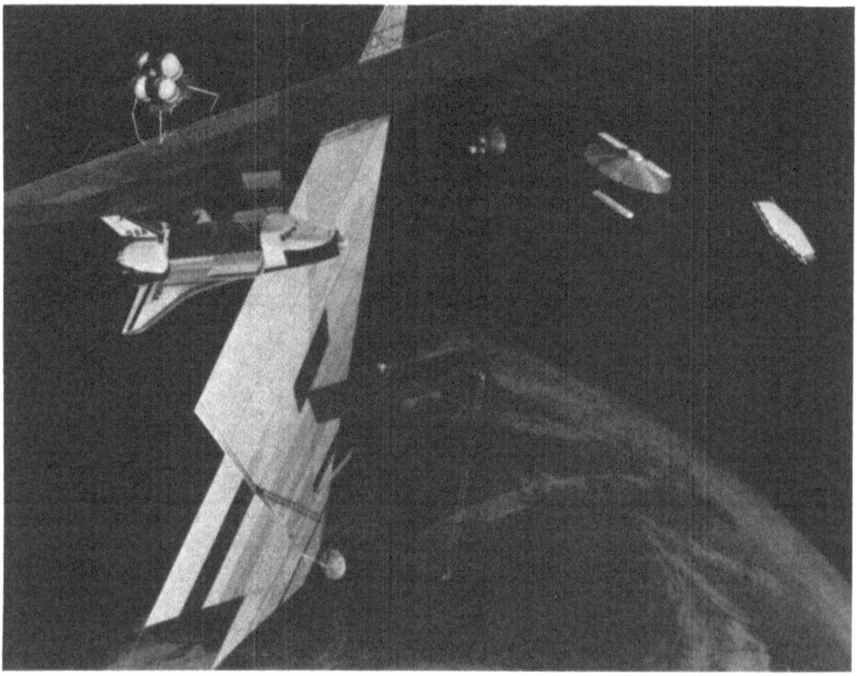

Figure 7. Large space systems require robot and automation technology for fabrication, assembly and construction in space.

Figure 8. Large space antennas are erected with the help of a space-based construction platform. The Shuttle brings the structural elements to the platform, where automatic manipulator modules under remote control perform the assembly.

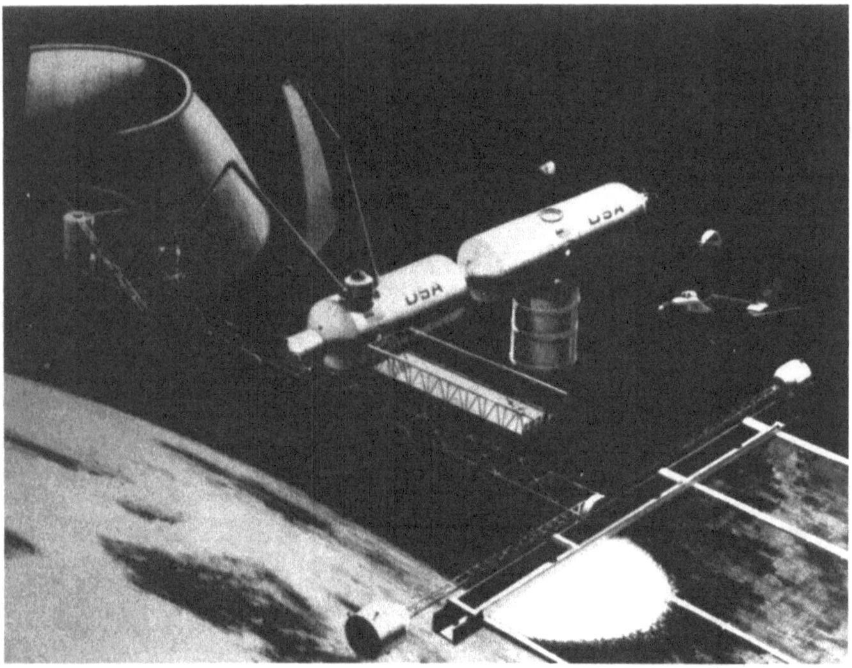

Figure 9. Construction of a space station. Bulk material is brought by the Shuttle. Structural elements are fabricated at the construction facility and then assembled by remotely controlled manipulators.

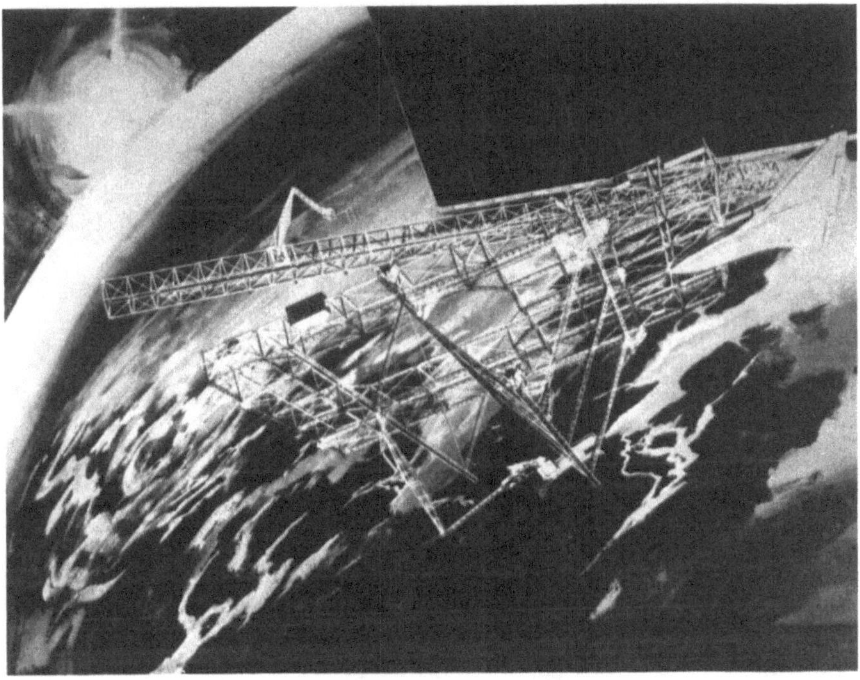

Figure 10. Complex construction facility in space with automatic beam builders, cranes, manipulators, etc., is served by the Shuttle.

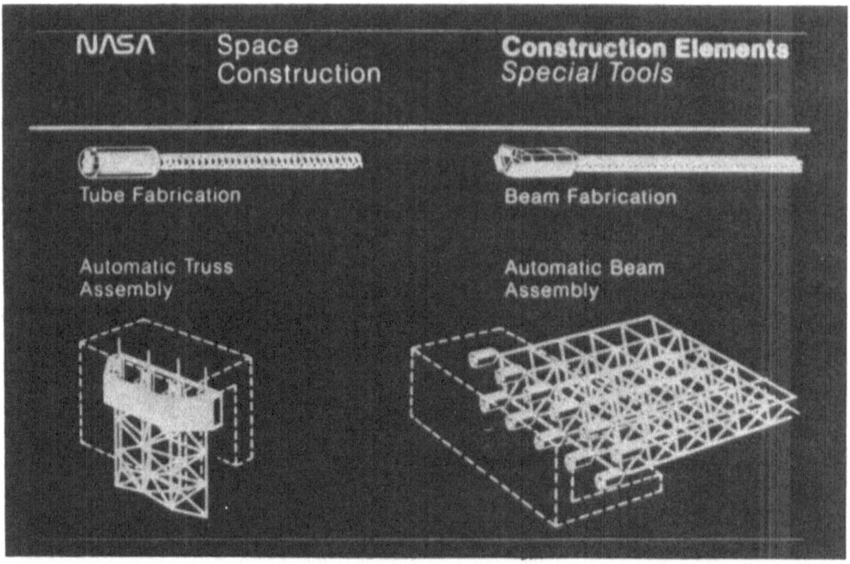

Figure 11. Automatic beam builders use sheet metal or bulk composite materials supplied by the Shuttle.

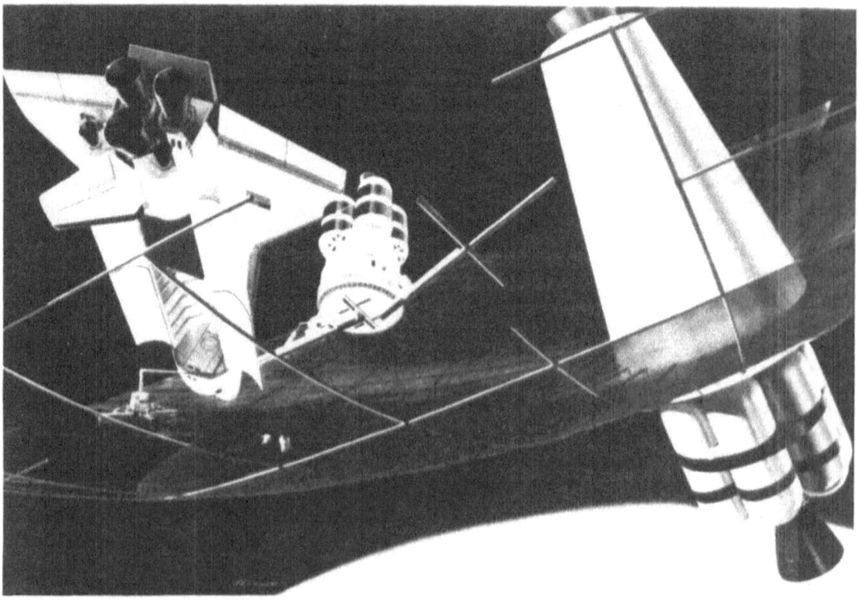

Figure 12. Space construction of large antenna systems with automated tools, teleoperated manipulators, and free-flying robots.

by robots attached to the structure. The functions which such a robot should be able to perform include calibration, checkout, data retrieval, resupply, maintenance, repair, replacement of parts, cargo and crew transfer, and recovery of spacecraft.

During and after construction, there should be a robot on stand-by for rescue operations. An astronaut drifting into space could be brought back by a free-flying robot. Such devices could also be on stand-by alert on the ground. The delivery systems for these rescue robots need not be man-rated. They can deliver expendable life support systems or encapsulate the astronaut in a life support environment for return to a shuttle, space station, or earth. They could also perform first-aid functions.

Another phase of space industrialization calls for a lunar base. After a lunar surface survey with robot (rover) vehicles, an automated precursor processor system could be placed on the moon. This system would collect solar energy and use it in experimental, automated physical/chemical processes for extracting volatiles, oxygen, metals and glass from lunar soil delivered by automated rovers (Fig. 13). The products would be stored, slowly building up stockpiles in preparation for a lunar base. The lunar base would be constructed using automated equipment and robots similarly as in earth orbit. After construction, general-purpose robot devices would be necessary for maintenance and repair operations. In addition, the lunar base would use all types of industrial automation (qualified for operation in space) that is generally used on earth for similar tasks.

In addition to acting as an enabling catalyst for some missions and reducing the cost of assembly and construction in space for others, the primary purpose of developing and applying robot and automation technology is the anticipated reduction of operation costs for all missions. It is, therefore, of interest to examine the NASA expenditures in various areas and to estimate potential cost savings, should an aggressive robot and automation research and development program be initiated. In a recent JPL study, it has been estimated that

Figure 13. Automated material processors on the lunar surface are serviced by robot vehicles with raw lunar soil.

about 600 million of the NASA budget in 1978 dollars could be saved annually by the year 2000 for ground operations, orbital operations, and data analysis. This does not include additional savings for incorporating advanced automation and machine intelligence techniques in design and testing operations. These projected savings also do not allow for the necessary expenditures to develop the required technology. Rough estimates for such investments range between 200 and 250 million in 1978 dollars to be spent on research and development between 1980 and 1995. This would be on the average about 0.4 percent of the NASA budget or about 12 percent of the OAST space budget. Although the estimated potential savings are subject to error, they indicate that space robot and automation technology is addressing high-leverage areas within the space program with large possible returns on the investment.

Conclusions

In summary, the space program is entering an era in which tremendous advances in sensor and electronics technology offer great opportunities to introduce advanced autonomous operations leading to large cost savings and mission capability enhancements. The possibility of applying robot and automation technology exists in almost every activity within the space program. It is anticipated that the space program will continue to be pursued with vigor during the coming decades, and it is therefore expected that the necessary tools will be available and that the opportunities afforded by electronics technology will be seized by the national space efforts in order to develop the required robot and automation technology that will facilitate a more capable and cost-effective space program.

References

1. Johnsen, E.G., and Corliss, W.R., Teleoperators and Human Augmentation, NASA SP-5047, December 1967.

2. Heer, E., "Remotely Manned Systems for Operation and Exploration in Space," First CISM-IFT.MM Symposium on Theory and Practice of Robots and Manipulators, 5-8 September 1973, Springer Verlag, Udine, 1974.

3. Deutsch, S., ed., "Teleoperator/Robot Development Task Team Report to the Acting Administrator, NASA," October 13, 1970.

4. Heer, E., "New Luster for Space Robots and Automation," Astronautics and Aeronautics, September, 1978.

5. Outlook for Space, NASA SP-386, January 1976.

6. OAST Space Theme Workshop, Final Report, Langley Research Center, Hampton, VA, April, 1976.

7. Kurzhals, P.R., "New Directions in Space Electronics," Astronautics and Aeronautics, February, 1977.

Suggested Readings

Key Economic Areas in Chinese History as Revealed in the Development of Public Works for Water Control, Ch'ao-Ting Chi, Paragon, New York, 1963.

Science and Civilization in China, Joseph Needham et al., Cambridge University Press, 1971 (esp. Vol. 4, Part III, Civil Engineering & Nautics).

The Pyramids of Egypt, I.E.S. Edwards, Penguin, Baltimore, 1961 (revised ed.).

Technology in the Ancient World, Henry Hodges, Alfred A. Knopf, New York, 1970.

The Ancient Engineers, L. Sprague De Camp, Doubleday, New York, 1963.

The Roads That Led to Rome, Victor W. von Hagen, World, New York, 1967.

A Social History of Engineering, W.H.G. Armytage, Faber & Faber, London, 1961.

A Thousand Years of London Bridge, C.W. Shepherd, John Baker, London, 1971.

The Cathedral Builders, Jean Gimpel, Grove, New York, 1961.

Sir Hugh Myddelton, Entrepreneur and Engineer, J.W. Gough, Oxford, 1964.

Edit Du Roi Pour La Construction D'un Canal De Communication Des Deux Mers, Océane & Méditerranée, St. Germain-en-Lay, 1666.

Lives of the Engineers, Samuel Smiles, John Murray, London, 1861.

Isambard Kingdon Brunel, L.T.C. Rolt, Longmans, London, 1957.

The Great Iron Ship, James Dugan, Harper, New York, 1953.

World Ditch, The Making of the Suez Canal, John Marlowe, Macmillan, New York, 1964.

The First Transatlantic Cable, Adele Gutman Nathan, Random House, New York, 1959.

The Nile Reservoir Dam at Assuan and After, Sir William Willcocks, E&FN Spon, Ltd., London, 1903.

Public Works, Vol. I, July-October, 1903, The St. Bride's Press, London.

The Path Between the Seas, David McCullough, Simon & Schuster, New York, 1974.

Of Dikes and Windmills, Peter Spier, Doubleday, New York, 1969.

The Tallest Tower, Eiffel and the Bell Epoque, Joseph Harriss, Houghton Mifflin, Boston, 1975.

The Bridge, Gay Talese, Harper & Row, New York, 1964.

The Big E, Edward P. Stafford, Random House, New York, 1962.

Cape to Cairo, Mark Strage, Harcourt Brace, New York, 1973.

The Great Iron Trail, The Story of the First Transcontinental Railroad, Robert W. Howard, Bonanza Books, New York, 1962.

The Impossible Railway, The Building of the Canadian Pacific, Pierre Berton, Alfred A. Knopf, New York, 1972.

The St. Lawrence Seaway, William R. Willoughby, University of Wisconsin Press, 1961.

A History of Dams, Norman Smith, Citadel Press, New Jersey, 1972.

The Great Wall of France, The Triumph of the Maginot Line, Putnam's, New York, 1961.

A Hole in the Bottom of the Sea, Willard Bascom, Doubleday, New York, 1961.

The Tunnel Under the Channel, Thomas Whiteside, Rupert Hart-Davis, London, 1962.

Supership, Noel Mostert, Alfred A. Knopf, New York, 1974.

The Hindenburg, Michael M. Mooney, Dodd Mead, New York, 1972.

The Ultimate Migration, R.H. Goddard, manuscript dated January 14, 1918, the Goddard Biblio Log, Friends of the Goddard Library, November 11, 1972.

Ethics and Bigness, edited by Harold Lasswell and Harlan Cleveland, Harper, New York, 1962.

Small is Beautiful, E.F. Schumacher, Harper & Row, New York, 1973.

The Spoils of Progress: Environmental Pollution in the Soviet Union, M.E. Goldman, M.I.T. Press, 1972.

A Wild Plan for South America's Wilds, "Fortune", December 1967, p. 148.

Global Earth-Shapers, "Fortune", April 1970, p. 78.

Managing the U.S. Supersonic Transport Program (A and B), ICH #9-675-049 and ICH #9-675-050; LIQUECOAL, ICH #9-678-046 (and other case studies available through the Intercollegiate Case Clearing House, Soldiers Field Road, Boston, MA 02163).

Power From the Sun: Its Future, Peter E. Glaser, "Science", Vol. 162, November 1968, pp. 857-886.

The Colonization of Space, G.K. O'Neill, "Physics Today", September 1974, pp. 32-42.

A Space Utilization Authority (SUA) Could be the TVA of the Future, Penelope G. Grodzka, Paper presented at the Conference: "Southeast 2001: The Next 25 Years," 12 and 13 November 1976, Urban Life Center of Georgia State University, Atlanta.

Implications of Low-Cost International Non-Voice Communications, Ithiel de Sola Pool and Arthur B. Corte, Report to the U.S. Department of Commerce, September, 1975.

"Limits to the Management of Large, Complex Systems," in Assessments of Future National and International Problem Areas, Vol. II, February 1977, Stanford Research Institute, Center for the Study of Social Policy, Menlo Park, California.

"Constraints on Large-Scale Technological Projects," in Assessment of Future National and International Problem Areas, Vol. II, February 1977, Stanford Research Institute, Center for the Study of Social Policy, Menlo Park, California.

"The Apollo Tradition: An Object Lesson for the Management of Large-Scale Technological Endeavors", by Robert C. Seamans Jr., and Frederick I. Ordway. In Interdisciplinary Science Reviews, Vol. 2, No. 4, pp. 270-305, December 1977, Spectrum House, London.

The Polaris System Development, Harvey Sapolsky, Harvard University Press, 1972.

Ptolemaic Alexandria, P.M. Fraser, Oxford, 1972. (three volumes).

The Emerging Japanese Superstate, Herman Kahn, Prentice-Hall, New Jersey, 1970.

"The Tragedy of the Commons", Science, Vol. 162, pp. 1243-1248, December 13, 1968.

Decision Support Systems, Peter G.W. Keen and Michael S. Scott Morton, Addison Wesley, Reading, Mass., 1978.

The New Management, William H. Gruber and John S. Niles, McGraw Hill, New York, 1976.

The Peckham Experiment, Innes H. Pearse and Lucy H. Crocker, Yale University Press, 1945.

Macro-Engineering and the Infra-Structure of Tomorrow, edited by Frank P. Davidson, L.J. Giacoletto and Robert Salkeld, AAAS Selected Symposium, Westview Press, Boulder, Colorado, 1978.

Index